FUNDAMENTALS OF AUDIO PRODUCTION

DREW O. McDANIEL

Ohio University

RICK C. SHRIVER

Ohio University

KENNETH R. COLLINS

Ohio University

PEARSON

Boston   New York   San Francisco
Mexico City   Montreal   Toronto   London   Madrid   Munich   Paris
Hong Kong   Singapore   Tokyo   Cape Town   Sydney

Editor in Chief: *Karon Bowers*
Series Editor: *Jeanne Zalesky*
Series Editorial Assistant: *Brian Mickelson*
Marketing Manager: *Suzan Czajkowski*
Production Supervisor: *Beth Houston*
Editorial Production Service: *Publishers' Design and Production Services, Inc.*
Composition Buyer: *Linda Cox*
Manufacturing Buyer: *JoAnne Sweeney*
Electronic Composition: *Publishers' Design and Production Services, Inc.*
Cover Administrator: *Joel Gendron*

For related titles and support materials, visit our online catalog at www.ablongman.com.

Between the time website information is gathered and then published, it is not unusual for some sites to have closed. Also, the transcription of URLs can result in typographical errors. The publisher would appreciate notification where these errors occur so that they may be corrected in subsequent editions.

Library of Congress Cataloging-in-Publication Data
McDaniel, Drew O.
  Fundamentals of audio production / Drew O. McDaniel, Rick C. Shriver, and Kenneth R. Collins.
    p. cm.
  Includes bibliographical references and index.
  ISBN-13: 978-0-205-46233-9 (alk. paper)
  ISBN-10: 0-205-46233-2 (alk. paper)
  1. Sound—Recording and reproducing. I. Shriver, Rick C. II. Collins, Kenneth Ray. III. Title.

TK7881.4.M453 2008
621.389'3—dc22   2007019716

Printed in the United States of America

10 9 8 7 6 5 4 3 2 1   11 10 09 08 07

Photo Credits: All photos courtesy of the authors, with the exception of **pp. 15, 16, 18,** Photo courtesy of Auralex Acoustics, Inc.; **p. 95,** Sennheiser USA; **p. 210,** Rycote microphone windshields; **p. 222,** Courtesy of Nagravision S.A., Kudelski Group.

CONTENTS

The authors approached this text as an opportunity to reconsider the way that people learn to become audio production professionals. Such a goal is complex, involving multiple factors that fall within what educational theorists define as cognitive and affective domains of learning. Cognitive learning concerns the acquisition of information, and can be broadly understood as a memorization process that includes visual and auditory as well as factual details. Affective learning is connected to a learner's frame of mind about a subject, and involves such things as attitudes, interests, and general emotional state. We have attempted to organize this text in a way that optimizes learning of subjects that producers need to master, especially complex technical topics that are important in the audio field.

To accomplish this goal the text is divided into three main parts—Theory, Tools, and Production Settings and Techniques. The three sections are labeled The Theory of Sound and Its Electronic Representation; Sound Tools and Applications; and Production Settings and Techniques.

In the first part—Theory—the reader is led through the basic physics of audio, both acoustic and electronic, including how acoustic energy propagates and how it is captured and manipulated. The authors believe that underlying principles of sound must be understood before moving on to explore how those principles are applied in practical situations.

The second part is devoted to the tools used in audio production, that is to say the actual equipment that producers employ to create their products. Based on theories laid out in the text's first section, chapters in this section present explanations of the design, construction, and uses of production tools such as microphones and mixers. We maintain that no one can be proficient in the use of production tools without a grasp of the fundamental theories that guide operation of audio production equipment.

The text's third part is devoted to presentations on the actual use of equipment in different production settings. The way that any piece of gear is put to use is normally dependent on its production context. For instance, microphones are used differently in a studio than in an outdoor setting, and the issues confronting a producer are directly related to the sort of work being done. Moreover, any equipment's use must first take into account its technical functions as detailed in the text's second section.

Much of the content of this text deals with technical discussions of different types of audio equipment and theories of their operation. There are two reasons why so much emphasis is placed on technical explanations. First, in the audio production field a producer frequently must also act as engineer, and therefore must be prepared not only to operate equipment but to set up facilities and troubleshoot technical problems as well. Indeed, in many production settings, there is little distinction made between the two roles. Having a firm understanding of acoustic and electronic principles will enable producers to troubleshoot problems, or to find a solution or a work-around that will get a production project back up and running. Second, it is the belief of the authors that knowledge of the technology makes even those who do not act as engineers better as producers. If producers understand technical factors that govern the operation of equipment, they will undoubtedly be more knowledgeable in using the tools of audio production.

CHANGE IS CONSTANT

In the field of audio, as in any technological field, the only thing that remains constant is change. As in many technological fields, the pace of change has accelerated rapidly over recent decades and this has occurred in sound technology too. An analysis of developments in the recording industry reveals that the interval between invention and commercial diffusion has traditionally taken several years. That is to say, it normally would take years for new technologies to saturate the marketplace, and subsequent improvements in technology to be developed gradually over a period of years. Figure P.1 plots significant innovations in audio technology through the late nineteenth and twentieth centuries. This figure suggests a trend in which the number of significant technological developments per decade has risen.

In the current environment, improvements in devices and software enter the marketplace with great rapidity. The quickened pace of technological development has brought about revolutionary changes in recording, storage, transmission, and distribution of audio, and new developments seem to arise on a daily basis. At the same time, established technologies can meet their demise just as abruptly. Some experts propose that the rapid pace of technological change has been speeded by the deregulation of telecommunications that has taken place in many countries around the globe.

LEARN TO LEARN

So, while it took over thirty-five years for the principles of magnetic recording to evolve from experimental concept to commercial production, contemporary products' life spans may be only a few years. For example, Philips' Digital Compact Cassette was introduced in 1992, lived a short, ten-year life, and is now considered to be defunct. Contemporary audio software applications are sometimes updated several times per year. Thus, it is probable that the most important thing producers can learn in this field is how to learn, because they will be doing it the rest of their lives. The phrase "lifelong learning" has become a popular catchphrase in the education field, and nowhere is that term more applicable than in a technologically rooted field like audio.

Timeline of audio recording technology

1870–79
- Edison tin foil cylinder. 1877

1880–89
- Edison electric phonograph. 1888
- Berliner patents flat disk. 1887

1890–99
- Poulsen wire recorder. 1898
- Marconi wireless U.S. to Italy. 1895

1900–09
- DeForrest tube amplifier. 1907

1910–19
- Scully disk lathe. 1917
- Armstrong superheterodyne. 1916
- Edison kinetophone. 1913
- Armstrong regenerative circuit. 1912

1920–29
- Nyquist digital theorem. 1929
- Neuman condenser mics. 1928
- Vitaphone disks sync with film. 1927
- Oxide coated paper tape. 1926
- 78 RPM electrically recorded. 1925
- RCA ribbon microphone. 1925
- First broadcast KDKA. 1921

1930–39
- First FM Broadcast. 1939
- Column loudspeaker. 1938
- RCA bi-directional ribbon mic. 1938
- Single element cardioid mic. 1938
- Condenser mics. 1935
- BASF plastic coated tape. 1935
- RCA cardioid ribbon mic. 1933
- Blumlein patents stereo. 1931

1940–49
- McIntosh amplifier. 1949
- Ampex 300 tape recorder. 1949
- RCA 45 RPM single. 1949
- Portable tape recorder. 1948
- Acetate tape by Scotch. 1948
- Columbia Long Play records. 1948
- Professional tape recorder. 1946
- Home wire recorders. 1946
- Altec Coaxial loudspeaker. 1943
- Stereo tape recordings. 1942
- RCA reference monitors. 1942

1950–59
- Nagra battery field recorder. 1958
- First stereo records. 1958
- First 8 track multitrack recording. 1956
- Ampex "Sel-sync" overdubbing. 1955
- Commercial stereo tapes. 1954
- Sony pocket transistor radio. 1954
- Plate reverb. 1954
- Ampex four track 35mm film. 1953
- Tone control circuit. 1952
- Bell labs introduces transistor. 1951
- Pultec active equalizer. 1951
- Les Paul creates Multitrack. 1950

1960–69
- Experiments with digital recording. 1969
- Time delay spectrometry. 1967
- Moog synthesizer. 1965
- Dolby type A. 1965
- Electret microphone. 1963
- Philips compact cassette. 1963
- SMPTE standards set. 1962
- FM stereo standards. 1961

1970–79
- Sony open reel digital recorder. 1978
- dbx noise reduction. 1978
- JVC introduces VHS. 1976
- 16-bit digital recording. 1976
- Sony Betamax. 1975
- Digital reverb. 1975
- Digital Stereo in studios. 1975
- CrO_2 Cassettes. 1974
- Quadraphonic recordings. 1972
- PCM stereo recording on U-matic. 1971
- Sony U-matic recorder. 1970
- Lexicon digital delay. 1970

1980–89
- Digidesign "Sound Tools." 1987
- R-DAT introduced. 1986
- First digital consoles. 1986
- Dolby SR. 1985
- Apple Macintosh. 1984
- Digital audio over fiber optics. 1983
- First CD players. 1982
- VCRs for digital recording. 1982
- Philips Compact Disk. 1981
- Sony "Walkman." 1980
- Hard disk recorders. 1980
- Digital multitrack. 1980

1990–99
- MP3 Players introduced. 1998
- DVD introduced. 1997
- 24 bit, 96KHz digital. 1996
- JAZ and ZIP drives. 1995
- Prosumer digital console. 1994
- Mackie "prosumer" 8-bus mixer. 1993
- Distant recording by ISDN. 1993
- Sony MiniDisk. 1992
- Philips DCC. 1992
- Quick Time released. 1991
- Alesis A-DAT. 1991
- Recordable CDs. 1990
- Dolby 5 channel surround. 1990
- ISDN links studios. 1990

In light of this, the authors have tried to impart an underlying and lasting conceptual understanding of capturing, manipulating, storing, and transporting sounds. We have provided ample historical background for the processes and techniques of audio recording, but we have attempted to minimize any treatment of topics that are peculiar to specific pieces of proprietary equipment or software.

It is our contention that when one understands something about how and why sound behaves as it does, and the reasons equipment operates as it does, one will be better equipped to control sound, to manage audio signals, and to effectively operate equipment to achieve creative objectives.

LISTEN

Although it is easy to become enamored of gadgets that are part of the production tool kit, the most important tools for the audio professional are one's ears. Becoming a critical listener is a vital first step to success in the audio field. There is no secret formula that will unfailingly produce great sound; for example, there is no single "right way" to place a microphone that will provide perfect recordings every time. Instead, producers must learn to listen. Each microphone will color sound in unique ways. Moving a microphone by a few inches, changing its direction, equalization, and gain can dramatically alter its sound. Thoughtful microphone placement based on the spot from which sound emanates and the way acoustic waves move is a starting point. Through careful experimentation and listening, skillful producers can arrive at a combination of microphones and placements that make it possible to capture sound exactly as intended.

WHAT IT ALL MEANS

One thing can be predicted—in the time it takes a reader to finish this text, some forms of audio technology will have changed significantly, possibly even becoming obsolete. Even so, our goal is that the learning acquired through reading this book will persist well beyond the fleeting lives of current technology.

ACKNOWLEDGMENTS

We would like to thank those who provided comments on the manuscript: Joseph Blaney, Illinois State University; Neil W. Goldstein, Montgomery County Community College; James Harley, Minnesota State University, Moorhead; Sam Lovato, Colorado State University, Pueblo West; and C. Collin Pillow, Arkansas State University.

CHAPTER ONE

SOUND AND ITS NATURE

Most audio books begin with a discussion of sound: what it is and how it moves under different conditions. The authors of this text believe that understanding the nature of sound is fundamental to the processes of controlling, capturing, storing, manipulating, and reproducing audio properly.

Students of audio sometimes struggle to comprehend the physics of sound, whether represented acoustically or electrically. For many, the easiest under-standing of such concepts comes from visualization; if one can picture an idea in the mind's eye, it can be readily understood. Toward that end, many authors begin their discussions by invoking the visual metaphor of the sound **wave**, typ-ically depicted as the familiar **sinusoidal**, or **sine**, wave (Figure 1.1).

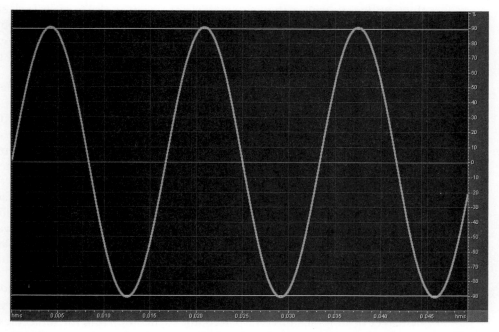

FIGURE 1.1　An audio sine wave displayed on the screen of an audio editor, created by an integrated oscillator.

Frequently, the sound wave is likened to the waves in water that emanate outward when a pebble is tossed into a pool, or the wavelike manner in which a stretched spring transmits a vibration. And while either of these similes offers some insight as to how sound travels in space, neither really describes adequately what occurs when sound is produced in an enclosed, three-dimensional space.

SOUND FROM MOLECULAR MOTION

This discussion will emphasize an alternative view of sound waves. The basis of sound is **vibration**. Whether those vibrations emanate from the vocal folds, the strings of a violin or piano, the reed of a saxophone, or the head of a drum makes no practical difference. Furthermore, sound is transmitted, or travels, as a kind of vibration. A proper view of the behavior of sound in the enclosed space of a studio, for example, rests on visualizing molecular vibrations and how they emanate from a sound source. Sound moves through the air by setting up variations in the density of air **molecules** that constitute vibrations. Air in a room is composed of molecules that are actually spread about sparsely, with space between individual molecules. These molecules are constantly in motion under the influence of minute temperature changes, wind currents, gravity, and other forces. If molecules were enlarged so that they were visible to the naked eye, one would see that even in a quiet, empty room, molecules are in nonstop motion, as if the air were fluid. Physicists explain that we are all immersed in an ocean of air molecules. And while those air molecules do have mass, and at sea level exert considerable weight on us—amounting to 14.7 pounds per square inch—we are not conscious of this. People literally swim through this sea of air molecules unmindful of the effort required.

If a person were to speak in this room filled with air molecules, the pressure of the air waves emanating from that person's voice would be enough to move the molecules nearby. These molecules would bump into the molecules next to them, and those molecules would bump into those next them, and so on, until finally the air molecules adjacent to the listener's ear would receive this vibration and exert pressure on the **tympanic membrane** (eardrum). This phenomenon is illustrated in Figure 1.2.

A more detailed explanation may help clarify the process of molecular displacement as sound travels in air. When a person speaks, as just described, some air molecules around the speaker are compressed. The compression wave of air molecules moves away from the speaker by being pushed into the next layer of air molecules. Naturally, this is a three-dimensional process taking place in a three-dimensional space.

To oversimplify somewhat, slight changes in air pressure caused by speaking move air molecules immediately surrounding the speaker. As noted, when this first layer of air molecules is displaced, they will bump into the molecules adjacent to them, thus causing them to be displaced. But as pressure returns to normal, the first layer of air molecules returns to its original place. This "bouncing back," or re-

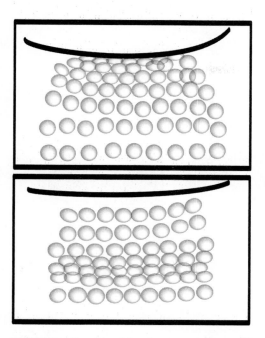

FIGURE 1.2   Air molecules are compressed by sound pressure. Compression waves are transferred through air molecules as they are forced to squeeze together with adjacent molecules.

turning, of the air molecules produces a sharp reduction in their compression. This condition is called **rarefaction**.

Therefore, the transmission of sound through air takes place by means of the slight movement of molecules. Air molecules vibrate in place, bouncing back and forth, as tiny changes in air pressure cause them to first become compressed together, then to rarefy by spreading apart. It is from these progressive molecular movements, emanating outward from a source, that the notion of a sound wave can be applied. The vibrations move in a wavelike fashion, as molecules ever farther from the source bounce against each other, as suggested in Figure 1.3.

Here is where the ripples-in-a-pool comparison may be instructive. As a person walks through a room, the body's movement creates waves and eddies in the air mass, as it would in a pool of water. If one imagines living in an ocean of air that acts in a fluidlike manner, it should be possible to visualize the waves that wash through this imaginary ocean. When a person speaks, vibrations of the vocal folds create regular variations in air pressure that radiate outward in waves in all directions, just like ripples in a pool.

The expenditure of **energy** is another factor that plays a key role. Energy is required to do work. Because air molecules have mass, as they are compressed, energy is spent in moving them. This part of the process is referred to as the

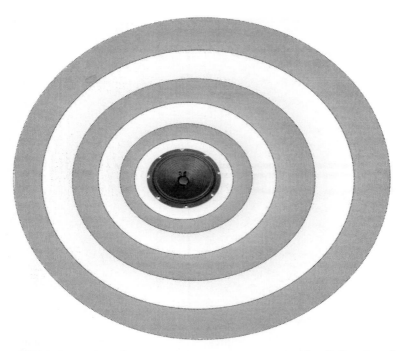

FIGURE 1.3   Sound pressure waves emanate outward in all directions from a sound source.

positive phase because there is a positive expenditure of energy. When the air molecules return, or rarefy, some of that energy is recovered, and this is called the **negative phase**. When molecules return to their original positions, a complete **cycle** has occurred: from stasis through the positive phase, through the negative phase, and back to stasis. Ultimately, once waves pass, the molecules will come to rest somewhere near their original starting points. The parallels between alternating pressure in a sound wave and the up-and-down movement of rippling water should be apparent. It is descriptive not only of how sound travels, but also of how energy is expended and recovered (Figure 1.4).

PITCH AND FREQUENCY IN SOUND

The **frequency** at which cycles of compression and rarefaction occur determines a sound's **pitch**. The fewer waves that occur in a period of time, the lower will be the perceived pitch of a sound and, conversely, the more rapidly vibrations occur, the higher the pitch. The standard unit of measure is the number of cycles per second or **hertz** (abbreviated as **Hz**). If a string on a piano were struck, causing it to vibrate at 440 times per second, the pitch of the resultant note would be 440 cy-

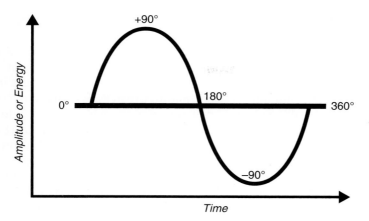

FIGURE 1.4   Phases of a single complete sound wave cycle.

cles per second or, more properly, 440 Hz. The frequency of 440 Hz describes the number of times the sound wave goes through cycles of positive and negative phases in the period of one second.

In its perception of changes in frequency, the human ear is imperfect. The range of frequencies that a typical listener can detect is thought of as roughly 20 Hz to 20,000 Hz, but this can vary widely from person to person, and commonly changes with age. In fact, the individual frequency response range is quite irregular, as the shape and folds of the ear alter the reception of various frequency sounds. In general, the human ear affects the perceived frequency response range so that an individual can most easily hear **mid-range** frequencies, which are those within which sounds of the human voice fall. This characteristic will be discussed in greater detail later, in what is called the principle of equal loudness.

Sounds of frequencies that are below 20 Hz are termed **subsonic**, or below those which can be heard. These low frequencies are often said to be felt rather than heard. Frequencies above 20,000 Hz (also abbreviated 20 **kilohertz**, or 20 **khz**) are **ultrasonic**, or above the range of frequencies that humans can hear. Sounds at the lower end of the human hearing range are termed **bass** frequencies and those at the upper end are called **treble** frequencies.

Frequency is also directly related to **wavelength**. Pressure waves that ripple through the molecules of air exhibit crests and valleys, just like waves in an ocean do. The distance from one crest to the next can be measured in meters or feet. A sound's wavelength is the distance between its peaks in density. Frequency and wavelength have an inverse relationship. As the frequency measured by the number of vibration cycles per second increases, the length of the waves will decrease. Awareness of this relationship is critically important in designing studio spaces and equipment, as will be shown later in this chapter. The key point is this: the lower the frequency, the longer the wave, and the higher the frequency, the shorter the wave.

SOUND AMPLITUDE

The quantity of energy that is used in creating a sound will determine the magnitude of the molecular displacement. Sound engineers refer to this sound magnitude as **amplitude**. This can be thought of as the number of molecules in motion or, more precisely, the amount of their motion. Greater molecular movement—or greater amplitude—will result in higher sound volume or louder sound. For the listener, changes in amplitude are perceived as changes in loudness. The greater the energy transferred to the tympanic membrane, the greater the air pressure on the eardrum and the louder the sound apparent to the listener. Acousticians often use the term **sound pressure level** to quantify amplitude. Changes in loudness are actually changes in the air pressure produced by sound waves.

A metaphor that might help in visualizing the amplitude phenomenon is the kinetic toy known as Newton's cradle, shown in Figure 1.5. When one ball is put in motion, one ball will bounce away on the opposite end of the cradle. Put two balls in motion, and two balls will be displaced, and so on. Regardless of the number of balls that are put in motion at one end, that same number will be displaced

FIGURE 1.5   The desktop novelty called "Newton's Cradle" helps illustrate molecular displacement produced by sound pressure waves. As the amplitude of a wave is increased, the magnitude of molecular displacement increases correspondingly.

on the opposite end. In the same way, the magnitude of molecular displacement near a sound source will proportionately displace the molecules at a distance, although the displacement will be attenuated because sound is dispersed as it moves away from the source.

In addition, owing to the mass of the molecules, energy is consumed in the processes of **attenuation**. The farther air molecules are from the source, the less energy there is available to move them. This is because the number of molecules that need to be moved increases as the sound wave travels farther and farther, causing a decline in the energy available. The result: amplitude declines as the listener moves away from the sound source. Move sufficiently far from the sound source and there will no longer be enough energy to move air molecules to produce perceptible sound.

In the fields of audio and **acoustics**, the standard unit of measure of amplitude is the **decibel**, commonly abbreviated as **dB**. The decibel, which is defined as one-tenth of a **Bel**, is sometimes identified as the smallest change in loudness that a trained ear can detect. In reality, most listeners are unable to detect changes in loudness smaller than 3 dB. Unlike the hertz, which is an absolute measure, acousticians customarily use the decibel as a relative measure, in describing differences in loudness. For instance, if a sound is boosted 6 dB, the sound after boosting is 6 dB louder than before.

The human ear does not have an unlimited capacity to respond to sound amplitude. Excessive volume can permanently damage the delicate mechanisms in the ear that transmit the small molecular vibrations through the auditory canal. This is a fact that audio producers should always be conscious of: excessively high sound pressure levels impacting the tympanic membranes can severely damage hearing. The effects of reduced air pressure on ears are well known to anyone who has flown on an airplane. In some ways, high-amplitude sound levels have similar effects. As the airplane rises, pressure levels drop, causing passengers to sense the change as pressure in their ears, sometimes quite painfully. The **dynamic range** of the human ear is generally considered to be about 120 dB. For this reason, 120 dB is often called the **threshold of pain**, and this is the amplitude at which irreparable or permanent damage begins to occur. Concertgoers sometimes experience a ringing in the ears after exposure to high sound pressure levels, an indication that the sound pressure levels were at or near the threshold of pain.

SOUND WAVE PROPAGATION

The phrase *speed of sound* is a commonplace expression. In the jargon used by acoustic specialists and sound engineers, the term **velocity** is often used to describe the speed at which sound waves travel. Use of this term carries with it the risk of confusing the principles that underlie the way that sound behaves because, as has been emphasized, sound does not really travel. Instead, waves of varying densities of air molecules move through layers of adjacent molecules as they respond to the energy supplied by sound sources.

Nevertheless, there is a definite and measurable speed at which these vibrations are communicated, and it is possible to determine with accuracy just how long it takes for the vibrations at the sound source to reach some distant reception point, say, one's ears. Because the ambient density of air molecules varies slightly with altitude, the velocity at which sound travels varies a little with altitude as well. For the sake of simplicity, acousticians normally consider the speed of sound at sea level—approximately 1,130 feet per second—sufficiently accurate to be used in making calculations. This figure is called a constant, even though minor variations occur as barometric pressure and altitude change. This amounts to roughly 770 miles per hour, which is what aeronautical engineers and pilots call **Mach 1.0** speed.

SOUND TIMBRE

Sounds that are produced by musical instruments are quite unlike the pure note of the sine wave mentioned earlier. The sounds of musical instruments, like most natural sounds, tend to be highly complex. By complex, it is meant that such sounds contain overtones and harmonics that color a sound. Overtones and harmonics give a sound its characteristic **timbre** or tone color. These overtones and harmonics are additional frequencies that accompany the **fundamental** or **dominant frequency** of a sound. In the previous example, a piano note of 440 Hz was mentioned, a note known as concert A. The sound of this note, when played on a piano, contains not only the 440 hertz fundamental frequency but also **harmonics** made up of multiples and submultiples of 440 Hz. The frequency of 440 Hz is dominant, but the note will also contain harmonics at 220 Hz (one-half the dominant, or the second subharmonic), 880 Hz (double the dominant), 1,320 Hz (three times the dominant), 1,760 Hz (four times the dominant), and so on.

Overtones result from frequencies whose interplay produces tones at the sum or difference of those component frequencies. For example, 440 Hz and 220 Hz might interact to create an overtone at 660 Hz. Harmonics and overtones occur at much lower sound pressure levels than the dominant frequency, but are a critical part of the piano's timbre, and their interactions may produce overtones of their own.

A change in frequency that is either one-half or double the dominant frequency is known as an **octave**. Thus, 880 Hz is one octave above 440 Hz (concert A). A note at 220 Hz is one octave lower than concert A. Similarly, 1,760 Hz would be one octave above 880 Hz, or two octaves above concert A. This relationship, the octave, is one that most listeners can detect as something special. Even persons who are musically untrained recognize that the two notes are somehow related and congruous, even if they cannot describe their relationship. The concept of octaves as a way of describing relationships among frequencies comes into play often in the fields of audio production and acoustics.

PSYCHOACOUSTICS AND THE PRINCIPLE OF EQUAL LOUDNESS

The objective study of how sound waves behave, especially in enclosed spaces such as rooms and halls, is known as *acoustics*. The more subjective study of how sounds affect humans, or how human brains interpret sound, is called **psychoacoustics**. Currently, some areas of inquiry in psychoacoustics include how locations of sounds are detected, how pitch is heard and perceived, and how the ear separates complex auditory signals occurring simultaneously. Research in these areas has practical implications for certain aspects of the audio field, such as improving spatial imaging and managing digital file sizes.

Harvey Fletcher and Wilden Munson of Bell Laboratories conducted some of the pioneering research in psychoacoustics in the 1930s. Their studies led to the publication of what has become known as the **Fletcher-Munson curves**, a set of graphs that visually demonstrate the **principle of equal loudness**. Fletcher and Munson's work was refined and extended in 1956 by D. W. Robinson and R. S. Dadson, and subsequently defined by the International Standards Organization in 2003 in ISO 226. Despite the later work, most audio practitioners refer to the graphs as the Fletcher-Munson curves (Figure 1.6). The principle of equal loudness as represented in the Fletcher-Munson curves describes the human ear's frequency response curve, as standardized through a series of widely administered listening tests. Acoustic specialists know through these tests that the human ear is

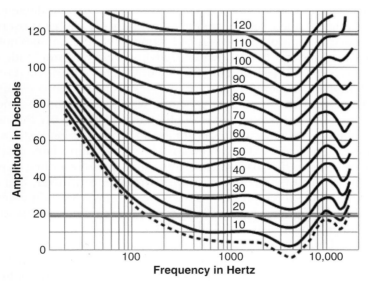

FIGURE 1.6   The Fletcher-Munson curves illustrate the principle of equal loudness, which suggests that mid-range frequencies are perceived by the human ear as louder than low and high frequencies.

most responsive to frequencies between 1 kHz and 6 kHz. They also know that a human's ability to hear frequencies below 500 Hz falls off steeply. The ability to hear frequencies above 7 kHz falls off too, but somewhat less sharply.

The Fletcher-Munson curves are referred to as equal loudness curves or contours because they illustrate that in order for listeners to perceive all frequencies as equally loud, the amplitude of frequencies outside the mid-ranges, roughly 1 kHz to 6 kHz, must be increased. In other words, listeners must boost the bass and treble portions of the audio spectrum to make those frequencies appear equally as loud as mid-range frequencies.

FREQUENCY AND TEMPORAL MASKING

Another imperfection of human hearing is its inability to distinguish and process the whole harmonic content of a highly complicated auditory signal. There can be so many frequencies present in complex sounds that the ear tends to distinguish only dominant ones. This is because dominant frequencies tend to mask weaker frequencies. This well-known and frequently studied phenomenon plays an important role in certain techniques used in the audio field.

Two kinds of **masking** have been identified: *frequency masking* and *temporal masking*. **Frequency masking** occurs because of the ear's limited ability to simultaneously process sounds of multiple pitches. To oversimplify slightly, a loud, 1,000 Hz, tone will have a tendency to mask a softer, 500 Hz, tone, so that the ear will not detect the lower frequency. Frequency masking is fundamental to the processing of digital audio and in designing transducers and converters for low bandwidth circuits and storage. Because it is known that the ear does not respond to all the frequencies in a full bandwidth signal, some lower-amplitude elements of the signal can be removed, and the ear will not perceive the loss. Masking simplifies digital audio file compression. A compression **algorithm** eliminates the low-level sound components in order to shrink file sizes. This can be done because engineers know listeners would not really hear them anyway. By this means it is possible to reduce data rates in digital storage and transmission systems, making file sizes more manageable. More on this will be provided in a later chapter.

Masking can also hide background noise in an audio recording or public address system. Analog tape hiss, thermal noise from an **amplifier**, and electrical noise interference are often present but disappear from perception when a moderately loud sound begins. Listeners hear only the intended signal because their hearing no longer discerns the masked noise sounds.

Temporal masking refers to the phenomenon in which the human ear has difficulty hearing a soft sound that follows a loud sound closely in time. Research on this subject has shown that the ear's recovery time is typically about five milliseconds. Moreover, strange is it may seem, pre-masking causes the ear not to hear a soft sound that immediately precedes a loud sound. This knowledge, too, is applied to the development of data conversion schemes in order to reduce the size of data streams.

Another application of research into the masking phenomenon is noise masking. This research has shown that **noise** is a powerful masker. The ear/brain will focus on the white noise to the exclusion of other low-level sounds; for example, conversations from adjacent office spaces. Some sound companies have developed noise-masking systems, which are purported to foster greater privacy through the use of either **white noise** or **pink noise**. These noises resemble the sound of rushing wind or water. Noise-masking systems usually involve a series of speakers placed above suspended ceiling systems. The speakers produce relatively low-level noise that is evenly distributed across the office area.

Still another area of psychoacoustics research is the study of how individuals **localize** sounds in space. The brain constantly makes calculations based on the short delays between the times when sounds arrive at the left and right ears. A sound originating to the left of the listener arrives at the left ear a few microseconds before arriving at the right ear, partly because the sound has to travel a slightly longer path, and partly as a result of the shadowing effect of the head. The brain takes that very short delay, termed the **interaural time difference**, into account when locating the sound to the left of the listener. The brain also factors in the relative amplitudes of the sounds, termed the **interaural level difference**, as the shadowing effect of the head will also lower the amplitude of the more distant sound and introduces some comb filtering. Finally, a frequency-filtering scheme compares the frequency content of the sounds arriving at the two ears.

The body of research that has grown out of the investigation of sound localization suggests that the frequency and the complexity of the sound, meaning its harmonic content, play key roles in the ear's ability to locate the sound source. Generally, very low frequencies and very high frequencies are more difficult to localize. Research into spatial localization has its practical application in the development of surround sound and virtual reality audio. Knowing which cues the brain relies on to localize sound allows sound designers to prepare soundtracks that manipulate sounds so that they appear to originate from intended points in space.

ACOUSTICS

A knowledge of how sound travels allows acousticians to design environments that permit control over sound characteristics. For example, one of the first problems that must be solved in designing a studio space is keeping unwanted sounds out, and at the same time stopping sound from traveling out of the studio space or into adjacent studio spaces. As already noted, sound can be thought of as a product of molecular vibration, since acoustic energy can only be transmitted through such motions. Achieving **acoustic isolation** from this point of view simply becomes a matter of stopping those molecular movements. Once defined this way, acoustic isolation translates into a mechanical problem that can be solved using well-established methods.

The two simplest techniques for discouraging vibrations of air molecules are to dampen them with something of great **mass**, or to mechanically **decouple** adjacent surfaces so that they cannot transmit the vibrations. Understandably, materials and objects exhibiting great weight are difficult to move or vibrate. Imagine a gigantic boulder, something at least the size of a large house, and then imagine what would happen if a person tapped on one side of that boulder with a rubber mallet. It would be unlikely that someone on the other side of the boulder could feel the tapping. The mass of the boulder would dampen the vibrations completely. This is the principle behind creating studio walls that are constructed of thick cement, or concrete blocks that are filled with cement. The walls become so massive that they do not transmit sound vibrations well, and serve instead to block the passage of acoustic energy.

Creating such massive obstacles introduces considerable expense. So, acoustic isolation frequently requires a compromise between cost and acceptable attenuation. Another, somewhat less costly method for discouraging sound transmission involves disconnecting, or decoupling, adjacent surfaces. Studio designers sometimes describe this technique as building a room inside a room. Using this method, walls are constructed of double sets of staggered studs. The double studs effectively disconnect the inside walls from the exterior walls. The technique is made more effective by adding several layers of drywall (adding mass) to the studs and filling the wall cavity with acoustic insulation (Figure 1.7). A second floor is

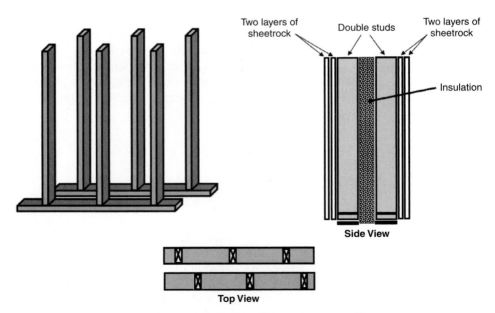

FIGURE 1.7   Acoustically decoupled studio walls are constructed by creating two separate walls that are physically disconnected from each other. To increase acoustic isolation, the cavity between the two walls can be filled with insulation material, such as fiberglass.

"floated" on top of the building's subfloor. The floating floor sits on spring-loaded or rubber shock absorbers that prevent vibrations in the studio from being transmitted through the floor to adjacent areas.

A second concern in designing studio space is managing the movements of sound within the room, or the reflection of sound waves within the enclosed space. This usually entails some type of **surface treatment**. A fundamental rule of studio design is to avoid parallel surfaces. This means that opposing walls should not be parallel to each other, and, ideally, the ceiling should not be parallel to the floor surface. This will discourage **room resonances** that can be created when **standing waves** are set up. Standing waves are sound waves that reflect off a surface in such a way that they either interfere with or add to later arriving waves. As a reflected wave arrives at the primary wave, it may be combined in a way that it is 180 degrees out of phase, which causes it to cancel out, or it can be 360 degrees out of phase, which reinforces the primary wave. Waves that are 180 degrees out of phase will be in opposition and will result in lowered volume at the resonant frequency, while waves that are 360 degrees out of phase will be in phase, and cause the resonant frequency to be louder than others (Figure 1.8). Such standing waves are set up when a sound wave bounces between two parallel surfaces. Standing waves are particularly problematic at the frequencies where the length of the wave is equal to the distance between the two surfaces, or is a multiple of that distance. For example, if a room is 20 feet long, a sound wave whose length is 20 feet long from crest to crest will set up standing waves. Similarly, wavelengths of 10 feet and 40 feet could also be an issue. So it is vitally

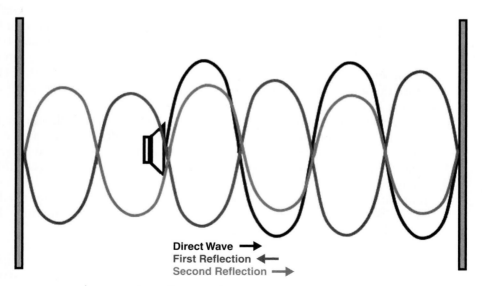

Direct Wave ➡
First Reflection ⬅
Second Reflection ➡

FIGURE 1.8   As sound is reflected within an enclosed space, standing waves may be created. Waves that are out of phase will cancel each other and not be heard, but those that are in phase will reinforce each other and thus sound louder.

important to design a studio space where length, width, and height of the room are not the same dimension or multiples of each other.

To understand this better, try to calculate the resonant frequency, or mode, for a room. To do this, first calculate wavelength. This is a relatively simple mathematical operation, described by the formula:

Wavelength = Velocity ÷ Frequency

Recall that sound velocity is a constant that at sea level is 1,130 feet per second. To solve for a 60 Hz frequency (an AC power hum, for example), simply divide 1,130 by 60, which will yield a result of 18.83 feet. A sound wave at 60 cycles per second would exhibit a length of 18.83 feet from crest to crest of the sound wave. By inference then, if the distance between studio walls were 18.83 feet, there would likely be a danger of setting up standing waves in the room when a 60 Hz sound filled it. The 60 Hz tone then would appear louder than other frequencies, or lower in volume than other tones, depending on the phase characteristics produced by the sound source. If the length and width are both 18.83 feet, the risk is magnified further. The following formula calculates the frequencies that might cause resonance problems in a given space:

Frequency = Velocity ÷ Length (of room dimension)

Using 1,130 feet per second as the speed of sound once more, the mode for a room that is 20 feet long can be determined by dividing 1,130 by 20. The resulting quotient, 56.5 Hz, is the frequency at which problems with resonance or phase interference can be predicted. Because the interference may occur at harmonics of the fundamental, one might also expect problems at 113 Hz, 226 Hz, 452 Hz, and so on. If this hypothetical room had a ceiling height of 10 feet, it is even more likely that a serious potential problem will exist (i.e., 1,130 ÷ 10 = 113 Hz). This is because the second harmonic of the room length (56.5 × 2), 113 Hz, is the same frequency as the fundamental of the room height.

Arranging studio walls so that they are not parallel will discourage these reflections, standing waves, and undesirable room modes. Additionally, careful attention to room dimensions during the design phase will help avoid other acoustic problems. It often happens, however, that the space to be used as a studio is already constructed and may be subject to problematic room resonances. Or the room may be too "live," or reflective of sounds, and yield excessive reverberation. In this case, the designer must turn to surface treatments to minimize those unwanted effects.

Surface treatments generally fall into three categories: *absorbers*, *resonators*, and *diffusers*. **Absorbers**, also sometimes called traps, are designed to capture and dissipate specific frequencies. One of the most common **acoustic treatment** absorbers is polyurethane **acoustic foam**. The foam is a spongelike porous material, with open cells that trap the movement of air molecules and thereby inhibit reflections and subsequent reverberation (Figure 1.9). Acoustic foams, like all such

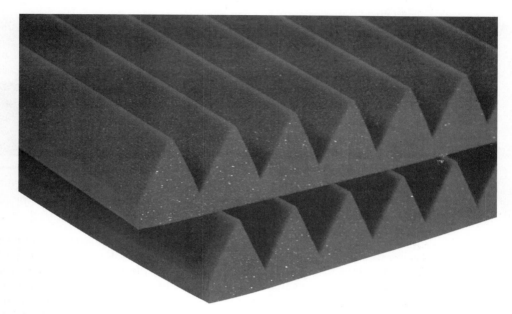

FIGURE 1.9   Acoustic foam is used for sound absorption because open cells in the foam trap and dissipate acoustic energy.

porous treatments having small surface openings, generally offer absorption only to frequencies in the upper part of the human hearing range. Thus, while acoustic foam wall treatment can help attenuate the high frequencies that contribute to a bright reverberant sound, it will do little to eliminate "boominess" at lower frequencies.

To absorb sound at the lower portion of the audio range, larger traps, sometimes called bass traps, may be needed. For many years, bass traps were designed on a model pioneered by Hermann Helmholtz. This device became known as the Helmholtz resonator. It is a chamber with a narrow throat, which resonates at a specified frequency, based on the chamber's volume. An easy way to envision how the Helmholtz resonator works is to picture what happens when blowing across the opening of a bottle partially filled with water. When liquid is drained from the bottle, the volume of air in the bottle increases and the frequency of resonance drops in pitch. The bottle has a resonant frequency that is predictable based on its volume.

To build a bass trap based on the Helmholtz concept, a large chamber with a small opening is constructed. In many control rooms, this is achieved by building a false wall in front of the rear wall. The false wall can be slotted to allow sound waves to enter the chamber between the walls. Varying the width of the slots permits a range of wavelengths to enter the chamber, as illustrated in Figure 1.10. The cavity is normally lined with absorptive material so that acoustic energy from low-frequency/longer waves trapped in the chamber will be dissipated. Rather

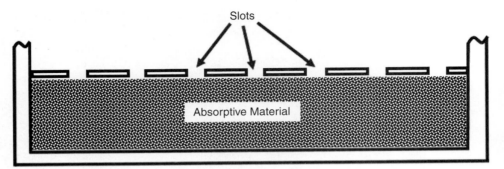

FIGURE 1.10 A bass trap is constructed of cavities that capture long wavelength, low frequency sounds and dissipate their energy in the absorptive material inside the cavity.

than resonating, the cavity catches and eliminates the lower frequencies. To be precise, this design should be called a Helmholtz absorber, rather than a resonator.

Nowadays, mass-produced sound control products are readily available to treat low-frequency resonance. Commercially produced bass traps are manufactured from polyurethane foam, which is denser than high-frequency foam, and is cut into shapes known as anechoic wedges. Placed in corners, these products can dramatically reduce low-frequency reflection (Figure 1.11).

Resonators also often find application in tuning listening environments. A room with small dimensions may not have sufficient volume to allow low-frequency waves to fully develop. As seen in the previous example, a room 20 feet

FIGURE 1.11 Commercially produced bass traps make use of anechoic wedge shapes to disperse and absorb low frequencies.

long will have a resonant frequency of about 56 Hz. Of course, frequencies lower than 56 Hz will exhibit even longer wavelengths. Those waves will not become fully developed in a room with a length of less than 20 feet. The result would be compromised low-frequency response below 56 Hz. To compensate, a series of resonators (with no absorptive material) can be installed to reinforce low fre-quencies.

Diffusers are used to inhibit reflections by breaking up parallel surfaces, and causing waves to scatter in different directions. This will reduce room resonances, and discourage other undesirable acoustic phenomena such as **flutter echoes** and excessive reverberation. Diffusers are often the first remedy applied to a problem room as they may reduce the need for absorbers. To achieve diffusion, the wall surfaces must be broken up by eliminating a single, large surface plane. One sim-ple method for creating diffusion is to apply strips or blocks of wood of varying thickness and width to the wall surface. This changes the topography of the wall from a flat plane to an uneven surface. A second type of diffuser incorporates hemispherical or cylindrical shapes attached to the wall surface, which will scat-ter the reflected sound waves in many directions, as illustrated in Figure 1.12. Some commercial products use absorbent tiles to achieve the same effect (Figure 1.13).

Trends in studio and control room design come and go. How surface treat-ments and acoustic design principles are applied change with the times as well. Be-fore the 1960s, the conventional wisdom was to make studios as acoustically **dead** an environment as possible, virtually eliminating any kind of reverberation. Stu-dios and control rooms were lined with absorptive materials such as thick drapes that left the rooms utterly devoid of reflected sound. But this type of environment is often uncomfortable for a listener. People are accustomed to receiving cues

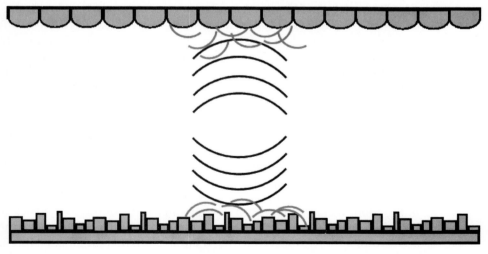

FIGURE 1.12   Irregular surface treatments can help eliminate standing waves by reflecting the waves at random angles.

FIGURE 1.13 Commercially produced absorbent diffusion tiles disperse waves at uneven angles and provide sound absorption.

about the size of the listening space by hearing sounds reflected from the walls, ceiling, floor, and other surfaces in the room. A completely absorptive space does not provide those cues.

For a time into the 1970s, control rooms were designed with the front (the end where the monitors are located) being more live, or reflective, and the back (behind the listener) being dead. The surfaces on which the monitors were placed were hard and flat—usually wood or stone—and were located adjacent to the large hard-surface plane of the windows. The back wall of the room incorporated a number of traps to discourage reflections. The idea was that sound reflected to the back of the listener's head would adversely affect the accuracy of the listening environment.

By the 1980s, one control room design had completely reversed that scenario. A design known as live end–dead end (LEDE) emerged. In LEDE design the front of the room was completely absorptive, to theoretically discourage any temporal smearing that might result from sound waves that were bouncing off the hard surfaces surrounding speakers before traveling to the listener's ears. The rear of the room was live, or reflective, to bounce some of the sound back to listener's back. But the LEDE design fell from favor just as quickly, and contemporary trends are toward a more moderately reverberant space that enhances a listener's ability to localize sound sources accurately.

SUMMARY

Sound exists as vibrations, whether originating in the vocal folds of the larynx as in speech, the strings of a piano when struck by the hammer, the reed of a woodwind as air is forced through, or the head of a drum when struck by a stick. The molecular activity of sound waves can be extremely intricate, with movements occurring at various frequencies and amplitudes.

Knowing this, the reader already has an idea of the best way to regulate the behavior of sound within an enclosed space; one must control vibrations. This is the basis for a science called acoustics, the name given to the study of the behavior of sound. It is this understanding of molecular movement that enables audio engineers to make devices that convert these vibrations into electronic signals and to make other devices that convert electronic signals back into sound. Ultimately too, this knowledge provides an understanding of how listeners hear or interpret sounds, a field of inquiry known as psychoacoustics. These and many other topics will be explored in subsequent chapters of this book.

CHAPTER TWO

CAPTURING SOUND ELECTRONICALLY

It is helpful to think of a sound or recording system as a chain, with each individual piece of equipment, each cable, and each connector serving as a link in that chain. The signal chain analogy is useful when an operator troubleshoots a problem. Beginning at one end of the chain and working through it in a process of elimination, technicians can isolate a broken link. Of course, as the old adage goes, "a chain is only as strong as its weakest link." If this is true, then no single piece of equipment can be more important than any another in shaping the quality of a recording or output signal. Even so, producers commonly believe that any audio chain's most critical link is its very first one—the one where an electronic audio signal is created.

The first step in recording or reinforcing sound is to convert the physical motion of air molecules, their **acoustic energy**, into electronic energy. This is accomplished by use of a device that falls into a class known as **transducers**. There are many kinds of transducers capable of converting energy from one form into another; for instance, electronic to physical, magnetic to electronic, electronic to magnetic, electronic to optical, and so on. However, the following discussion will focus exclusively on transducers that capture sound—those that convert acoustic energy into electrical energy.

MAGNETIC AND ELECTRONIC PRINCIPLES

Transducers used in capturing sound make use of familiar physical phenomena that generate electronic voltages. As early as 1873, James Clerk Maxwell theorized that coils of wire and a magnet could interact with one another. Today, this relationship is often demonstrated in elementary science classes by connecting a dry cell to a coil of wire. The current flowing through the coil will cause a deflection of a compass needle or other permanent magnet brought close to it. This results from a magnetic energy field surrounding the coil's wire that is produced by the electrical current in the coil. The magnetic field is aligned in a uniform manner of **polarization** according to the current's direction of flow. Once subatomic particles known as electrons are set in motion by force of the battery, a magnetic field arises. This electronic movement is the basis of **electromagnetic** physics.

Another classroom demonstration may be set up using a permanent magnet and a coil of wire. It is possible to induce current to flow in the wire by moving a permanent magnet along the coil. The action of a moving magnetic field of the permanent magnet forces electrons into motion throughout the coil's length. To fully understand this phenomenon, the basics of atomic physics and of electronics must be considered. Atoms are mainly made up of three types of smaller, sub-atomic particles: positively charged (+) **protons,** found in an atom's nucleus; neutrally or non-charged (0) **neutrons,** also in the atom's nucleus; and, orbiting around the nucleus, rings of negatively (–) charged **electrons.** Electronics derives its name from the study of electrons and their physical properties.

Most school children have learned that "opposites attract and likes repel," and this is precisely true in the field of electronics. Electrons are locked in their orbits by the balanced charges of the nucleus and the electron rings; the number of protons and electrons are normally equal. When the number of protons (+) and electrons (–) are equal, an atom remains in a stable state. However, some materials possess atoms arranged so that electrons in the outer rings are weakly locked in place. These electrons are known as **free electrons,** meaning that the electrons are free to be moved. Materials with free electrons are called **conductors;** the more free electrons, the better the conductor. Those free electrons can be dislodged from their nuclear orbits with comparative ease. For example, since like charges repel, it is possible to push electrons out of their atomic orbit by forcing electrons in.

A conductor such as a wire can be imagined as a string of atoms that have free electrons. If somehow electrons are forced into one end of the wire, an excess of electrons is created that will be pushed through to the other end of the wire by hopping from atom to atom. Current flows by a kind of chain reaction—an electron is forced into the outer orbit of an atom, causing another electron to jump out of that atom's orbit and to enter the outer orbit of an adjacent atom, thus again causing an electron to be forced out of the second atom's orbit, and on and on. This is how electrical current flows in any **circuit.** Electrons are pushed into the circuit from an electron source, and a current of electrons moves along from atom to atom. Of course this does not mean that electrons pile up at the end of a wire carrying a current. Because electrons repel each other, they will be equally spaced along the wire. See Figure 2.1.

This one-directional flow is descriptive only of the current flow in a **direct current (DC)** circuit, in which the force causing electron movement is often supplied by a battery. In DC sources, polarity never changes. The force required to sustain current flow is known as **electromotive force (EMF).** The force supplied by an ordinary household outlet, on the other hand, is **alternating current (AC),** and the polarity of electron movement alternates between positive and negative. Household current in the United States and North America reverses its direction, or its **phase** sixty times per second. The frequency of this EMF is 60 cycles per second. In most of the rest of the world, electrical current exhibits a

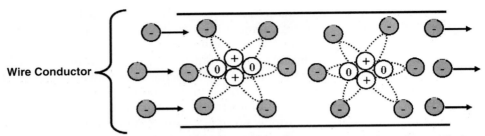

Wire Conductor

Electrons put into motion by electromotive force push other electrons through a conductor.

FIGURE 2.1   Electrical current flows in a conductor when free electrons jump from their atomic orbits to adjacent atoms, forcing other electrons from their orbits.

frequency of 50 cycles per second. Alternating current EMF is normally supplied by electrical generators. Generators are constructed of an arrangement of permanent magnets and coils that, when set in motion relative to one another, induce current to flow first in one direction then in the reverse direction. Any current flow requires two wires for a complete circuit or loop. In a DC circuit, one wire will be positive, known colloquially as "hot" or the plus side, and the other will be negative, called "cold" or the minus side.

DYNAMIC TRANSDUCERS

These simple points about the way current is induced to flow in a circuit guide the principles of many electromagnetic transducers. Through **induction** electromagnetic transducers create an EMF. One important type of electromagnetic transducer converts slight movements of air molecules into an electronic current. As just noted, if a permanent magnet is in close proximity to a coil, any movement of either one will cause an electromotive force to be set up in the coil. Transducers employed in capturing audio are often constructed so that a coil moves within a magnet's field. This arrangement is preferred because coils usually have less mass than magnets, so fixing the magnet in place and allowing the coil to move along with surrounding air molecules is more efficient in producing EMF output.

Microphones designed on these principles are called **dynamic microphones**. They operate by responding to changes in air pressure that result from sound waves. When air molecules are displaced by positive sound waves, the pressure wave pushes the microphone's diaphragm away. The diaphragm is constructed of a thin metallic or plastic membrane on the rear of which a small lightweight coil, called the voice coil, is attached. As shown in Figure 2.2, when the diaphragm moves in and out in response to air pressure variations, the coil moves within the magnet's field and a current flow is induced in the coil. The resulting EMF, its output voltage, depends on the amount of coil movement, which in turn depends on the sound waves' loudness.

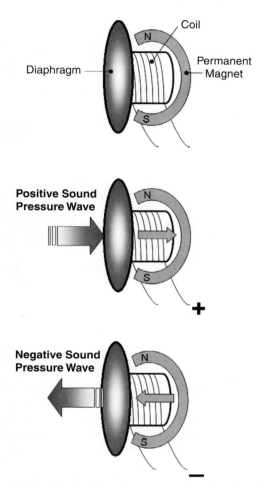

FIGURE 2.2   Electrical current flow is induced in a dynamic microphone when sound pressure waves cause a coil to move within the field of a permanent magnet.

When a pressure wave is positive, it pushes the diaphragm away causing a positive current flow and electrons move through the coil in one direction. When a pressure wave is in its negative or rarefaction phase, the diaphragm is pulled outward. During a negative phase electrons in the coil reverse their direction, and the EMF's polarity also reverses. Consequently, the electrical output of the transducer will be an analog of sound pressure waves impacting the diaphragm at any instant. The resulting output can be described as an alternating current at **audio frequencies**, a current that presents positive and negative phases.

Apart from microphones, there are additional devices used to convert sound energy into electrical energy that exploit the magnet-coil relationship. Pickups,

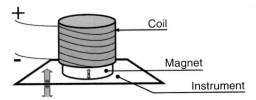

FIGURE 2.3   A musical-instrument electromagnetic pickup employs the same principle as the dynamic microphone. Vibrations in the instrument are transferred to the coil. As the coil moves with the field of the permanent magnet, current flows.

used on electric guitars, basses, and other steel-stringed instruments, are constructed of coils wrapped around permanent magnets. Vibrations of the instrument's steel strings create corresponding variations in the field of the magnet. This in turn induces an alternating current in the coil. A similar type of transducer using the same magnetic basis of operation is attached directly to the instrument's body (Figure 2.3). This type of transducer is used on stringed instruments that have gut or nylon strings and on percussion instruments, among others. Tiny vibrating movements in the instrument's body are transferred to a magnet in the transducer that vibrates sympathetically within a coil, thus producing an output current.

RIBBON TRANSDUCERS

Ribbon microphones make use of magnetic fields and their ability to generate current as well, although the way they work is a bit less intuitive. In the **ribbon microphone**, current is induced to flow along a thin strip of corrugated metal foil suspended within the magnet's field. The foil simultaneously serves as the microphone's diaphragm and the conductor for voltages developed within the magnetic field. Often two magnets are used in a ribbon microphone in order to concentrate their combined fields. These bar magnets are placed so that their powerful magnetic fields completely surround the ribbon (Figure 2.4).

As the foil in the ribbon microphone flexes responding to variations in air pressure caused by sound pressure waves, a small current flows along the ribbon. This current is proportional to the displacement of the ribbon within the magnetic field. Unlike most other design types, the ribbon microphone is inherently **bidirectional**. That is, it responds equally to sound pressure variations from either side of the ribbon, owing to the method by which the ribbon is suspended within the microphone.

CONDENSER TRANSDUCERS

A second common method for making an electronic analog of sound waves involves a different approach to energy conversion. This method is based on electro-

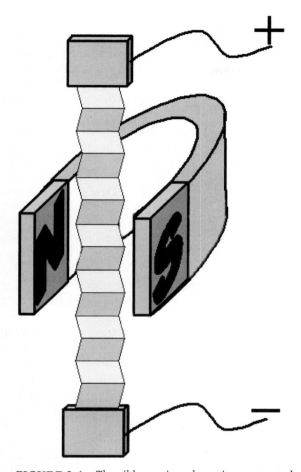

FIGURE 2.4   The ribbon microphone is constructed of a thin strip of corrugated metallic foil suspended within the field of a permanent magnet.

static or **capacitance** principles, rather than magnetic principles as in the dynamic and ribbon mics. Microphones utilizing this method are usually called **condenser microphones**, although they may be infrequently termed capacitance, **tube**, or *valve* microphones.

In order to electronically represent changes in air pressure, a condenser microphone takes a voltage applied to it from an external source, and then proportionally alters that voltage according to variations in air pressure striking the microphone's diaphragm. The external voltage is called **phantom power** (i.e., not visible to the user) and is external to the microphone.

An electronic component called a *capacitor*—also less often known by the outdated term *condenser*—is constructed from two conductive plates that are held in very close proximity to one another, but separated by a small nonconductive gap known as a **dielectric**. In a capacitor, the dielectric may be ceramic, mica, Mylar, or other plastic, or air or some other nonconductor. As electrons are supplied to one plate of a capacitor, some electrons will be forced to flow away from the adjacent plate, even though the two plates are not physically in contact. This is a function of the natural repulsion exhibited by particles of similar magnetic or electrical charge.

A condenser microphone makes use of this principle: the closer the two plates—in other words, the thinner the dielectric—the greater the displacement of electrons, and, conversely, the greater the separation between plates, the less electrons on one plate will affect the other. In the condenser microphone the front plate is a metallic diaphragm, and the backplate carries the current flow out of the microphone. The diaphragm is pulled tightly across an insulating ring, somewhat like a drumhead attaches to a drum. The ring is fixed to the metal backplate. The insulating ring separates the backplate and diaphragm by only a few thousandths of an inch, so that all that lies between them is a thin layer of air that serves as the dielectric of this capacitor, as shown in Figure 2.5. Phantom power voltage is applied to the diaphragm and backplate, usually on the order of 48 V. The voltage may be supplied via its cable by a power unit connected to the microphone, or by a source in a mixer.

As sound waves strike the microphone's diaphragm, its diaphragm moves slightly and this plus the varying density of air molecules causes tiny currents to move off and onto the diaphragm. The electron currents produced by this action are so small they must be amplified before use can be made of them. To accomplish this, condenser mics contain miniature preamplifiers powered by the same voltage sources used to charge the diaphragm. Once electrical signals have received preamplification there will be an output adequate for use with equipment such as mixers, recording equipment, and mic preamplifiers.

Similar to the electromagnetic dynamic transducer, condenser transducers can be adapted for use as an instrument pickup. By attaching the unit to an instrument's body, vibrations of the instrument will affect the distance between the backplate and diaphragm, just as they do in condenser microphones.

CARBON PARTICLE TRANSDUCERS

Although not really suitable for professional recording purposes, **carbon microphones** were once widely used for voice communications, including standard use in telephones, for many decades. Even today, carbon microphones are not uncommon in applications such as intercom headsets. The carbon microphone pickup is constructed from finely powdered carbon particles contained in a small cup. Carbon is somewhat conductive; the more compressed the carbon granules, the less resistance they exhibit to current flow. In these microphones, a current is set up through particles in the carbon cup. As compression of the particles changes

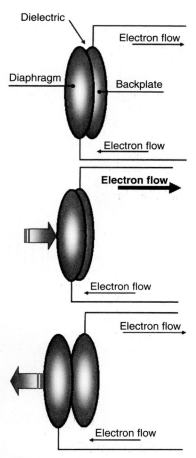

FIGURE 2.5   Sound pressure waves alter the distance between the two charged plates of a condenser microphone, which causes a current flow to be output.

through the action of a diaphragm pressing against the carbon granules, the current flowing will vary correspondingly (Figure 2.6).

As a result, the carbon microphone converts acoustic energy into electrical energy by altering a current, possibly supplied by a battery, flowing through the microphone. As a positive compression wave strikes the diaphragm, an attached piston compresses the carbon, thus reducing its resistance and producing an increased current flow. During the wave's negative phase or rarefaction, as the diaphragm is pulled away, the carbon particles are loosened and their conductivity drops, producing a decreased current flow. At times carbon transducers can accumulate a static electrical charge in the carbon cup. The static buildup results in an increase in noise output by the microphone and reduced audio output. A slight

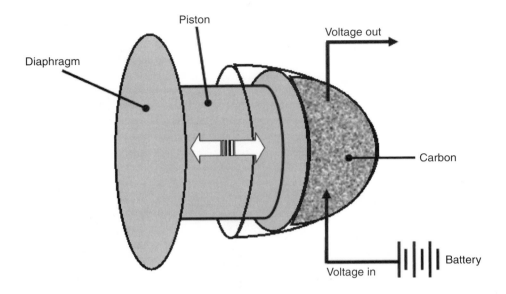

FIGURE 2.6   Granules of carbon conduct a current in a carbon microphone. As the carbon is compressed by sound pressure waves, the current flow increases.

but sharp rap on the microphone usually breaks up the tightly packed carbon particles and dissipates the static charge.

PIEZOELECTRIC TRANSDUCERS

A few microphones and music pickups make use of piezoelectric devices. These components take advantage of an unusual ability of certain crystalline minerals and ceramic materials to generate an electrical voltage when subjected to physical stresses. This is known as the *piezoelectric effect*. Whenever these crystals are stressed by the application of physical pressure, an EMF is produced at their surfaces. When the pressure is removed, the current ceases.

In a microphone or pickup, a thin slice of the piezoelectric material is fixed in place and attached to a diaphragm. Small conductive wires are attached to the crystal to carry away the current that it generates. As the diaphragm is moved by air pressure changes, it flexes the crystal (Figure 2.7). This produces an output voltage that matches changing air pressure. This scheme can be adapted for instrument transducer applications. For obvious reasons, mics using such devices are known as crystal or **piezoelectric microphones**. Crystal microphones are inexpensive to manufacture, but suffer from poor frequency response. They are, however, perennially popular with harmonica players, because their bandwidth is well suited to the instrument.

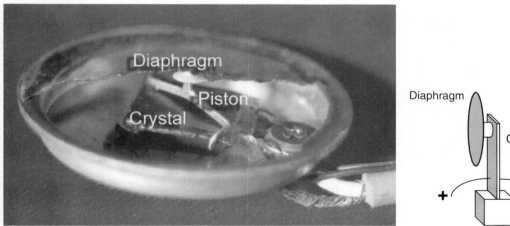
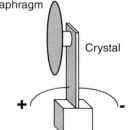

FIGURE 2.7   Crystal microphones make use of the ability of piezoelectric materials to create voltage when stressed by sound-pressure waves. The photo reveals the internal construction of a crystal pickup.

REVERSED TRANSDUCERS

There are endless variations on the kinds of microphones described here, all of which convert movements of air molecules or musical instrument bodies into an electronic representation of sounds. Interestingly, some transducers can operate in reverse, so that instead of converting acoustic energy into electrical energy, electrical energy is turned into acoustic energy. For example, anyone placing an order at a drive-through restaurant menu board actually communicates through a loudspeaker that is made to function as a microphone. Although the fidelity is not good, it hardly matters as long as the attendant can hear the orders accurately.

CHAPTER THREE

DIGITAL AUDIO

The ways of making an electronic representation of sound can be divided into two broad categories: analog and digital. Devices that produce analog signals are ones in which sound-pressure waves are translated into corresponding voltage levels. In devices like this, as air pressure rises or falls in sound waves, so does the voltage representing that sound wave—by an exactly proportional amount. **Digital** systems go one step further and translate the rising and falling voltage values into actual voltage measurements. To move analog audio into what is called the "digital domain," these continuous voltage waveforms must be subjected to rapid, regular measurements and the resulting values reported as a number.

By about 1990, consumer acceptance of the audio compact disk—CD—format had become so widespread that the entire music recording industry underwent a complete transformation, moving from analog to digital systems. This soon led to the same shift in other segments of the audio field. Digital audio technology had been employed in various ways for more than two decades, but a number of technical breakthroughs at the time led to a large leap forward. The advantages of digital technology are so significant that analog equipment today fills only a narrow range of applications in the audio field. This is true even though a few enthusiasts claim that analog technology produces a more pleasant sound quality. The benefits of using digital rather than analog methods will be outlined shortly, but first, let's look at how digital representation of sound is accomplished.

DIGITAL CODING

The manner in which voltages are converted into numeric readings is conceptually simple, but technically demanding. Humans live in an analog world; sound is made of continuously varying sound pressure waves, and our ears depend on those waves for the perception of sound. So digital processes can be used only as intermediate steps between the sound captured by a microphone and the rendering of sound by loudspeakers or headphones—all audio begins and ends in analog waveforms. When the analog audio is converted into digital—for example, following a microphone preamplifier—a device known as an **ADC** (**analog-to-digital converter**) is employed. An ADC makes periodic voltage measurements of the input signal at exactly timed intervals and then outputs those values as a number. Each value is termed a **sample**, because it is a measurement of the voltage waveform at a selected instant in time. At the other end of the chain, the digital signal must be restored to an analog format in order to make it into a sound again, and a **digital-to-analog converter** (**DAC**) is used for this purpose.

Numbers normally encountered in day-to-day life are of the base ten, recorded in a total of ten different numbers, 0 through 9. Numbers like this are referred to as decimal numbers. This is not the case in digital audio. For reasons of electronic simplicity, the numbers that stream out of the ADC are of a base two, using only two numbers, 0 and 1. Numbers of the base two are called **binary** numbers. Each digit of a binary number (either a 0 or 1) is called a **bit**, which is a shorthand expression for *b*inary dig*it*. Therefore, a 16-bit number would have 16 digits consisting of zeros and ones. In a 4-bit binary numbering system the decimal number 1 is equivalent to 0001, 2 is 0010, 3 is 0011, and so on. For a more comprehensive explanation of this numbering system, readers should refer to a mathematics text; a full explanation of the binary notational system is beyond the scope of this discussion. Fortunately, for producers there is seldom a need to worry about binary numbers because the hardware employed usually handles all computations without a need for human intervention.

Two aspects of the voltage measurements used in creating digital signals deserve careful attention: sampling frequency and the numbering scheme used to record voltage levels. Samples must be collected frequently if the rapid changes in voltage are to be faithfully reflected in the series of measurements. For obvious reasons, the more frequent the measurement, the more accurate will be the resulting trace of the analog signal. The minimum measurement frequency depends on the maximum frequency of the signal being reproduced digitally. This minimum sampling rate, often referred to as the **Nyquist frequency**, is twice the highest frequency produced in the signal. So, for example, to reproduce audio up to a frequency of 18,000 Hz, the sampling rate must be at least 36,000 times per second, or 36 kHz. However, this is a bare minimum; higher sampling rates can produce even greater sound accuracy.

Each sample consists of a binary number, the value of which is proportional to the voltage reading of the analog signal at that instant. As the voltage rises, the value of the binary number increases. The numbers reported in samples are of fixed length, commonly 16, 24, or 36 bits, or digits, long. A binary number like this is called a **word**. Longer words allow for greater precision in measurements, in the same way that grading on a 0 to 100 scale makes possible finer distinctions than a 1 to 10 scale. If samples are recorded in words of 8 bits' length, only 256 different values can be recorded—a cruder approximation than would be the case for a 16-bit word that can have 65,536 possible values.

On the surface, it would seem that for maximum quality a producer would prefer to use the highest possible sampling rate and greatest number of digits in each sample. But increasing both of these produces demands for higher data-processing capability and, ultimately, greater storage capacity for recordings. For example, use of 16-bit words requires more storage space and faster processing than 8-digit words. As a consequence, the **word length** chosen for any digital project inevitably will be a compromise between the higher quality of large numbers and the speed and compactness of shorter word lengths. Similarly, a doubling of sampling rates demands twice the processing speed and twice the storage capacity. In sound, samples of 12-bit numbers produce results acceptable for some purposes, but 16-bit audio clearly sounds better, even to an untrained ear. Frequency bandwidth, dynamic range, and **distortion** rates imposed by the conversion from analog to digital are largely determined by the word length of each sample and by the sampling rate. For this reason, there continues to be a gradual shift to word lengths of 20 bits or more in professional audio systems and to sampling rates higher than the conventional 44.1 kHz.

The problem with analog systems is that in passing signals through successive stages of electronic equipment, voltage changes through amplification must remain exactly proportional. It is impossible to create amplifiers that are capable of this kind of perfect accuracy, and so it is inevitable that slight imperfections creep into rising and falling voltage traces. Such errors tend to accumulate, so that the more devices a signal passes through, the greater the risk of errors. These errors are exhibited by greater **intermodulation** and harmonic distortion as well as increases in noise. The key advantage of digital technology lies in the simplification afforded when digits are represented electronically in a binary format. Digital devices interpret or produce signals that code the value of each bit in a "high" (on) or "low" (off) state, in which a high state represents a "1" and the low state stands for a "0." By convention, a signal level between 0 and 0.4 volts (V) will be interpreted as a 0, and 1 will be represented by a level between 2.4 and 5.0 V. Signal levels outside these ranges are treated as indeterminate. The same sorts of errors in voltage representation occur in digital circuits as in analog circuits, but they make no difference as long as discrepancies do not push voltages outside the broad "high" and "low" bands that represent the 1 and 0 binary digits.

As long as the numbers represented in each sample are accurate, there will be no distortion introduced by the minor voltage errors that occur in digital systems. For this reason, digital equipment can be expected to have superior technical specifications and superior sound qualities. Often in digital equipment, the actual distortion and noise is so low as to be difficult to measure. A digital system's dynamic range—the range between the noise floor and the maximum signal level, or **signal-to-noise ratio (SNR)**—depends on the word length employed. An 8-bit system has an SNR of 48 dB, and a 16-bit system has a potential SNR of about 96 dB. The theoretical dynamic range of other word-length systems can be roughly estimated by multiplying the word length by 6.

ANALOG-TO-DIGITAL CONVERSION

As noted, producers rarely have to concern themselves with the coding or decoding of digital signals at the level of binary numbers. Digital equipment uses circuits that manage conversion and processing of digital signals in a way that is transparent to equipment operators. Despite the advantages of digital systems, however, they are not without problems, and the producer should understand how to avoid them. On the whole, once audio has been converted into digital form, maintaining its integrity poses relatively few difficulties. Most errors occur in translation, either from analog to digital or from digital to analog. For this reason, producers customarily take great pains to avoid moving back and forth between analog and digital domains any more than absolutely necessary; once audio has been digitized, it is best to do all subsequent processing in the digital domain, converting back into analog only to listen to the material. The following discussion highlights potential problem areas in processes that convert audio signals from the analog to digital and digital to analog.

In moving analog audio to the digital domain, an immediate concern is the audio signal's **bandwidth**. By tradition, the audio spectrum is considered to be 20 Hz to 20,000 Hz. Even though this is actually slightly wider than typical human hearing, these figures have become an accepted standard, possibly because they are easily remembered. The problem occurs when signals are present that lie above the audio spectrum; that is, above 20,000 Hz. This can happen if the microphone is capable of generating output above that frequency, as some do, or it can happen if the analog circuits preceding the digital converter produce distortion products that fall above the audio spectrum. If analog audio is present that is higher in frequency than one-half the sampling frequency, the Nyquist sampling criterion will not be met, and errors in the voltage readings will surely result. This occurs because the sampling cannot adequately trace rapid changes in the voltage waves. Errors of this sort are evident in increased noise output.

To avoid errors resulting from audio content that exceeds the bandwidth appropriate for the sampling frequency, ADCs must incorporate an extremely sharp cutoff low-pass filter at the input of the A/D converter. This type of filter is termed an **anti-aliasing filter**. It passes signals below a specified frequency—in this case half of the Nyquist criterion—while simultaneously rejecting higher frequencies. Even if the audio signal is not likely to contain frequencies above half the sampling frequency, anti-aliasing filtering is customarily applied as a precaution.

A number of technical factors affect the ability of an ADC to process the conversion, especially those connected with the voltage measurements of the waveform. The section of the ADC that actually measures and records the voltage taken each time the audio is sampled is known as a quantizer and the process of measuring is called **quantizing**. At the heart of the quantizer is a "track-and-hold" circuit, sometimes known as the "sample-and-hold" circuit. It consists of a gate that opens briefly to admit the input voltage to a storage circuit. This voltage is maintained at a constant level until the gate reopens for another sample. The sample voltage is held between samples so that stable readings can be made. The quantizer takes each measurement, translates it into a scale appropriate for the word length of the sample, and then sends this value onward as a binary number. Speed of the measurement is critical, because numeric values must be determined in the interval between each sample's reading. A "successive approximation" method is often favored because it is capable of rendering measurements of one bit per clock pulse.

If this method is used by the ADC, the minimum clock frequency of the processing system would thus be equal to the word length multiplied by the sampling rate. Thus, for a sampling rate of 48 kHz and for 24-bit samples, the ADC would require at least a clock speed of 1.152 MHz, a rate easily obtained using current technology. Commonly employed audio sampling rates are 32, 44.1, 48, or 96 kHz. Using one of these industry standards ensures compatibility with other digital devices as the audio is processed and avoids the possibility that additional conversions will be required later, something that might possibly lead to degraded signal quality.

DIGITAL-TO-ANALOG CONVERSION

To complement the work of ADCs, digital-to-analog converters restore the original analog signal from a stream of digitally coded readings. DACs do this by reversing the coding process, constructing a continuous analog waveform proportional to the voltage measurements. One form of DAC creates a voltage pulse that matches the value of each sample. Another approach used in DACs is called the *1-bit dual digital-to-analog converter*. This kind of DAC uses a circuit called a *delta-sigma modulator*. This circuit creates an approximation of the original waveform by measuring average changes in the voltages from sample to sample. This kind of DAC must have a very high clock speed, and relies on the insensitivity of the ear to noise introduced by the conversion near the upper frequency end of the human hearing range.

In any case, the output of these circuits is only an approximation because it contains significant errors and frequency components not present in the original audio signal. Figure 3.1 illustrates the output of such a device. As the drawing reveals, even though the waveform is continuous, it does not vary smoothly; the waveform has a stairstep shape. The result is a "stepped approximation" of the original waveform. A low-pass filter can be used to eliminate the squared edges of the waveform, but this approach is seldom entirely satisfactory when used as the sole correction method. A low-pass filter usually cannot completely remove the unwanted audio components caused by the notched shape. The inadequate filtering can be detected as a "gritty" sound quality.

The errors that arise from the lack of a smooth waveform at the output of an ADC appear in the form of noise. **Oversampling** is a technique commonly applied to this problem. In oversampling, an ADC creates intermediate estimated samples to supplement the known sample values provided by the data stream. These estimates are derived by interpolation; that is, two successive values are compared and an additional value is inserted partway between the two known values. By this means, if two successive values had a decimal equivalent of 64 and 128, an intermediate value of 96 could be added between the two. The result is a doubling of data points used to construct the waveform and would be called *two-times oversampling*. Even more estimated intermediate points can be inserted, so that if samples were quadrupled, it would be termed *four-times oversampling*, and so

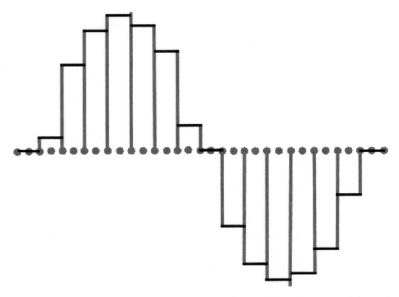

FIGURE 3.1   A digitized waveform is only an approximation of the original analog signal. It traces an irregular voltage pattern, resembling a stairstep.

on. As a rule, the gains of adding interpolated sample estimates diminish quickly and little additional improvement can be obtained beyond sixteen-times oversampling. Figure 3.2 shows how these additional data points improve the shape of the waveform. In effect, oversampling shifts the noise created by sampling errors into higher frequencies, which are less noticeable to the human ear and are easier for low-pass filters to remove.

DATA MANAGEMENT

The benefits afforded by digital technology—in the form of greater accuracy, lower distortion, and less noise—come with trade-offs. Foremost among these is the increase in bandwidth of digital signals. For instance, to transform 20 Hz to 20 kHz audio-to-digital signals based on 44.1 kHz, 16-bit samples requires a bandwidth of 16 times 44,100, or about 705 kHz. Digital audio in this form thus needs a bandwidth about thirty-five times greater than the same signals in analog form. This translates into requirements for a massive storage capability and rapid throughput. Most audio content for consumer use is stereo, so any data stream would need to have twice the throughput requirement as **monophonic** audio. For example, just one minute of stereo audio at 44.1 kHz recorded in 16-bit sample requires 10 megabytes (MB) of storage capacity. A **byte** is a set of data, often consisting of eight bits. As time has passed, processing-speed capabilities have met or exceeded needs as technology has advanced, but storage requirements have been an ongo-

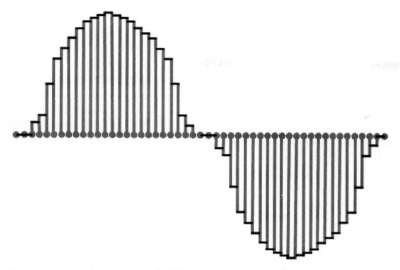

FIGURE 3.2   An oversampled digital waveform, although still only an approximation, presents a trace more like the original analog signal. Noise created by errors in the over-sampled waveform are much higher in frequency and thus easier to remove by filtering.

ing problem. Basically, two remedies have been available: (1) develop storage for-mats with greater capacity or (2) devise a means of shrinking the data stream to reduce throughput and storage requirements.

Data-storage capability of media available to audio professionals has ex-panded steadily over recent decades, in response to demands for higher-quality audio recordings that generate large masses of data. In the 1990s, the CD was the standard format used for mass-data storage. During that decade, the typical ca-pacity of hard drives used in workstations grew from 100 or 200 MB to 10 or 20 gigabytes (GB). At the time of this writing, data storage in the range of 1 or 2 ter-abytes (1,000 to 2,000 gigabytes) are commonly available. Similarly, the DVD has replaced the CD as a standard portable storage medium, and the DVD has evolved from a single-layer disk to a double-layer disk with twice the storage capacity. Other, ever-higher-density recording technologies are under development, so there is no sign the trend is slowing.

In spite of the successes of more and more storage capabilities, reduction of the data stream has important additional benefits, and this has led to develop-ments in the field of **data compression**. Digital data compression exploits the fact that any set of data is likely to be filled with repetitive samples. Perhaps the most familiar of the data reduction methods makes use of MP3 data compression. As is well known, MP3 recordings produce good-quality, if not outstanding-quality, sound recordings that require a fraction of the storage space needed by uncom-pressed audio. Satisfactory results can be obtained by using MP3 recording modes that reduce storage to less than 10 percent of the uncompressed files. The **MP3** method of compression is a format developed by the Motion Picture Experts Group (**MPEG**), an organization that formulated the standard that is now widely adopted throughout the world's audio and video industries. (MPEG formats for compression of digital video are widely used, too.) Another approach to audio-digital compression is used in the Sony-developed MiniDisc (MD) recording format designated as ATRAC. This method produces excellent audio quality for compressed digital recording and has been commonly utilized on MD recorders.

All audio-digital compression is based on data reduction of two types—eliminating redundant samples and eliminating audio components that will not be audible to the listener. In audio, a portion of the samples are bound to have identical values. If the number of these same readings is large, compression encoders collapse such samples into a single value in a way that decoders can reconstruct all original values on replay. This cuts throughput and shrinks storage and data-handling demands. Coding systems like MP3 also take into account the fact that human hearing is unable to detect all sounds in complex material, and therefore portions that won't be heard can be removed from the data without any noticeable result. For example, as discussed in an earlier chapter, it is known from psychoacoustics—the field that studies human hearing—that significant masking occurs when multiple frequencies are simultaneously present in sound. It is known that if one frequency rises above a certain level, it will make other frequencies inaudible through a covering-up effect. Under these conditions, masked frequencies can be eliminated from the data stream. Also, because human hearing is known to be most sensitive between 200 Hz and 5,000 Hz, and the hearing threshold drops substantially below and above this band of frequencies, some frequency components below the threshold of hearing outside this band can be dropped, too. Taking into account these physiological factors, MP3 and ATRAC can achieve enormous data reductions in digital data streams.

Analog technologies continue to have their rightful place in the audio field, and not merely as nostalgia pieces. Some analog equipment lends a subtle quality that many find pleasing to the ear. Ironically, this coloration can be achieved synthetically nowadays by the use of digital processing methods. So even if one lacks access to, say, an analog recorder, treating the digital signal with the proper processor or computer software can approximate its sound. Nevertheless, for all but a few purposes, digital audio is the preferred form employed to transport, process, and store sound in audio production.

CHAPTER FOUR

MICROPHONES

Microphones are perhaps the most critical part of the audio production chain. These devices act as gateways, translating sounds into electronic signals. If errors crop up at this stage they can never be fully corrected. As noted in an earlier chapter, microphones sense changes in air pressure produced by sound waves and convert them into an accurate voltage representation of sounds. Like loudspeakers, **microphones** are classified as transducers—devices that convert one energy form to another. Acoustic energy is taken in by microphones and changed into electrical energy. Producers should be aware that in any audio chain, transducers are the devices most likely to introduce inaccuracies.

Microphones can be classified in many ways—by the way they create an electronic output, by the directions from which they pick up sounds, by their intended uses, and by their ability to accurately reproduce different frequencies. All these factors must be taken into account when selecting a microphone for any application. Indeed, microphone choices are key factors in the success of a recording session. If it is a small session, the quality of sound captured will be determined largely by the microphones employed. But if the session is large and requires many microphones, the complexities multiply because each individual mic must be viewed as part of a whole interrelated set of audio sources.

The selection of an individual microphone depends mainly on a knowledge of its sound reproducing characteristics—in short, how voices and instruments sound when reproduced by a microphone. Ideally, the microphone should have a perfectly flat response, in other words, be able to reproduce all frequencies equally well between 20 Hz and 20,000 Hz. Of course, this cannot be expected of real mics. Microphones are imperfect devices, and each one has particular shortcomings that shape the way it sounds. If every microphone were capable of flat frequency response and had similar distortion characteristics, then all microphones would sound the same and selection would not be a concern. Since this is not the case, producers must choose the mic that best matches the qualities of the sounds being picked up. In most situations, these considerations are subtle, and the sound differences among very good mics are not dramatic. Even though differences among mics are slight, choosing the proper microphone is still important. A major part of a producer's set of skills is learning how to hear these subtleties and take advantage of them.

In addition to frequency response, dynamic range must be considered when choosing microphones. *Dynamic range* is the term used to describe the difference in level between the loudest and softest sounds that a microphone can reproduce. The dynamic range of a device is defined by its signal-to-noise ratio. A microphone's signal-to-noise ratio is the difference, measured in decibels, between the loudest and softest sounds it a can faithfully capture. The upper sound level limit of a microphone is determined by its overload characteristics. As sound levels increase, a point will be reached at which a microphone's output becomes unacceptably distorted because of its inability to cope with the high level of acoustic energy. Overloading occurs in a microphone when pressure waves cause its diaphragm to flex to the point where it can deflect no further. When this happens, sounds that it renders will become seriously degraded. The lower sound level limit is determined by the noise produced internally by a microphone's signal-generating components.

TECHNICAL FEATURES

Of the many types of microphones available, only three are in common use by audio professionals—the dynamic, condenser, and ribbon microphones. Of these three, dynamic mics are seen most often in a wide range of production applications. The dynamic microphone is rugged and extremely reliable, and many have good sound-reproducing characteristics. In addition, dynamic microphones are comparatively affordable.

Dynamic microphones depend on electromagnetic mechanisms to render electronic images of sound. Construction details of the dynamic transducer are described in Chapter 2, Capturing Sound Electronically. These mics have a diaphragm constructed of a thin metallic or plastic membrane on which a small, lightweight coil is attached. The diaphragm-coil assembly is placed within the field of a powerful permanent magnet. As sound waves strike the diaphragm, it moves the voice coil within the magnetic field, and this causes a voltage to be produced in the coil. Windscreens of plastic foam or fabric are usually employed to block gusting air movements that might cause noise or might damage the mic's diaphragm. Figure 4.1 shows a few examples of popular dynamic mics.

Even the most expensive dynamic mics tend to exhibit moderate levels of frequency distortion, however. That is, their response is not even, or "flat," across the audio spectrum. Dynamic microphones tend to have weak frequency response at frequencies above about 12,000 Hz. Not only does output usually drop at higher frequencies, but signal accuracy has a tendency to degrade as frequency rises.

The ribbon microphone operates on magnetic principles, too, somewhat like the dynamic mic. The main difference between them is that the diaphragm and coil assembly in the dynamic microphone is replaced by a metal foil strip in the ribbon mic. This kind of mic is often large and heavy, due to the massive ribbon and magnets commonly used in its construction. Adding to its bulk is the need for

FIGURE 4.1   Dynamic microphones, left to right: Stedman N-90, Shure SM-7, Shure SM-58, and Electro-Voice ND-468.

an internal transformer or preamplifier to boost the mic's output level and impedance to one that matches mixer and mic preamplifier inputs.

During the 1940s and 1950s, in the heyday of old-time radio, ribbon microphones were the microphone of choice, but their popularity waned afterward. Few of the older ribbon microphones accurately reproduced sound; most had limited high-frequency response characteristics. A graph of their output usually showed a highly irregular frequency-response curve. However, some music producers were fond of the rich harmonic content these mics generate, and they continued to use them for certain brass instruments and occasionally on vocals. In recent years, the ribbon microphone has enjoyed a resurgence of popularity. Few new designs had been introduced after 1960 until about 2000, when a new generation of ribbon designs became available. The new ribbon mics present somewhat flatter frequency-response characteristics, and most are more compact as compared to the classic types. There are even examples of ribbon mics that are meant to be handheld.

Although the ribbon mic is large, it is still a sensitive and somewhat delicate instrument. For this reason, ribbon mics should never be used outdoors because a gust of wind could permanently damage its ribbon. Figure 4.2 shows some examples of ribbon microphones.

The condenser microphone is mechanically simple, but the physics behind its operation is complex. Unlike the dynamic and ribbon mics, the condenser mic relies on the changing capacitance of the pickup element. Again, the construction of

FIGURE 4.2   Ribbon microphones, left to right: Oktava MJI-52-02 and Cascade Fathead.

a condenser mic is detailed in Chapter 2, Capturing Sound Electronically, and examples of this type of mic can be seen in the illustration in Figure 4.3.

The condenser microphone is characterized by a smooth and flat frequency response. Thanks to their internal preamplifiers, most condenser mics also produce high output levels. Because of their excellent sound-reproduction abilities, condenser mics are likely to be the preferred choice whenever audio accuracy is the most important consideration. Condenser mics also tend to have superior transient response capabilities—that is, the ability to handle percussive sounds such as drums, explosions, and gunshots. On the other hand, condenser microphones are sometimes less convenient to use than other types because of their need for external phantom power. The condenser mic is also prone to wind damage and to overload from high sound levels.

A variant of the standard condenser mic is the **electret condenser microphone**. These mics are just like other condenser microphones, except that the voltage applied to the diaphragm comes from an electret—a device that maintains a permanent electrical charge. Even though the electret charges the mic's diaphragm, it still requires power for its internal preamplifier, and this is usually

FIGURE 4.3   Condenser microphones, from left to right: Oktava MK-319, AKG-414, AKG C-2000B, AKG C-1000S.

supplied by penlight cells inside the microphone case. Most electret microphones are a bit noisy, but they do offer high output levels, satisfactory frequency response, and good reliability. Prices of electret condenser mics are usually modest, and even pricey ones typically cost less than a few hundred dollars.

PICKUP PATTERNS

Microphones are also classified according to their directional **pickup patterns**. Sounds arriving from different directions are not necessarily reproduced equally well by a microphone. Some mics take in sound only from a narrow angle in front, and others reject sounds from certain directions. Differences in pickup patterns are useful in alleviating problems that may arise in production settings. For instance, directional mics can be used to reject unwanted sounds, reduce sounds reflected from walls or nearby objects, reduce feedback, or isolate desired sound sources.

Pickup patterns are usually portrayed by graphs showing the relative sensitivity of microphones to sounds coming from different angles around the pickup point. These graphs are called polar diagrams. A polar diagram consists of a series of circles within circles, each one growing progressively larger as it moves away from the center. The circular figures display the sensitivity of a microphone at various angles. A **polar pattern** depicts the relative sensitivity of a mic to sounds arriving at angles from head-on (0 degrees) to the rear (180 degrees) to the front again (360 degrees) and all points in between, always with the greatest sensitivity

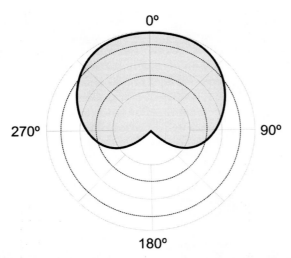

FIGURE 4.4   The polar pattern of a cardioid microphone, showing sound rejection from the rear.

at 0 degrees. The response at any angle can be determined by the number of circles the response curve reaches or exceeds. The distance between each circle indicates some fixed amount of output, typically 5 or 10 dB.

A useful directional pickup pattern is the **cardioid pattern**, so named because of the response outline's resemblance to a heart shape. The polar diagram in Figure 4.4 shows this cardioid outline. In the cardioid mic, the null, or point of least sensitivity, is at the rear of the mic at 180 degrees.

Microphones with the narrowest pickup pattern reduce sounds from their sides and rearward as well as the back, so that only the front of the mic picks up any significant sound. Known by various names, such as hypercardioid or supercardioid, these microphones employ a cardioid design modified so that a main lobe, or region of greatest sensitivity, has a narrow acceptance angle formed around 0 degrees. An example of a hypercardioid mic is shown in Figure 4.5. Note that the polar diagram reveals a number of nulls and minor lobes at the side and back of the mic. These are typical of this type of microphone. It might seem that directional mics could be useful in a great many applications, but most producers use them sparingly, and only when needed, because mics like this seldom have outstanding sound qualities. Highly directional microphones usually compromise accuracy and frequency flatness in order to achieve their narrow pickup angles.

Bidirectional mics are sometimes useful in professional settings. This kind of mic is sensitive to sounds from the front and back, but rejects sounds at the sides. The result is a **bidirectional** ("figure-8") **pattern** as shown in Figure 4.6. As can be seen, the polar pattern has lobes at the front (0 degrees) and the rear (180 degrees), with sharp nulls at the sides at 90 degrees and 270 degrees.

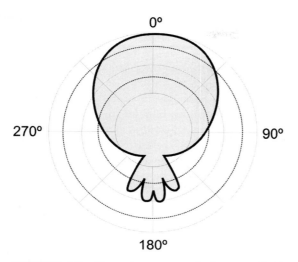

FIGURE 4.5   The polar pattern of a hypercardioid microphone, illustrating sound rejection from the sides and the rear.

Finally, the **omnidirectional microphone** picks up sound from all directions equally well. As can be seen in Figure 4.7, the omnidirectional polar pattern resembles a circle because this mic exhibits little variation in sensitivity to sounds arriving from any direction.

By and large, the method of creating an electrical output by a mic, whether dynamic, ribbon, or condenser, is unrelated to its directional pattern. A microphone,

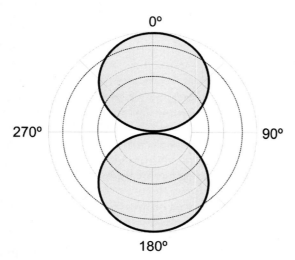

FIGURE 4.6   A bidirectional microphone polar pattern accepts signals from two sides.

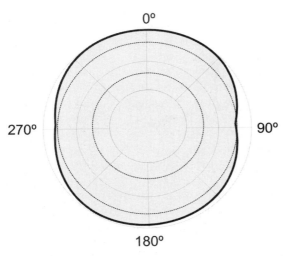

FIGURE 4.7   An omnidirectional microphone polar pattern shows approximately equal sensitivity to sound arriving from all directions.

regardless of its type, can be manufactured with any desired pickup pattern. Omnidirectional and cardioid mics are by far the most common in studio use, although hypercardioid mics are important in field recording. Producers usually choose a mic based on frequency response and the quality of sound reproduced by particular sound sources. Pickup pattern is a factor, but tends to be a secondary consideration in most production circumstances.

MICROPHONE USAGE

The impedance of a microphone refers to its ability to supply a real electrical current at its output. High-impedance microphones have high internal resistance and/or reactance and are able to produce only a very small output current. Low-impedance mics can provide larger electrical currents, albeit at lower voltages. Typical values for low-impedance mics range from less than 50 ohms up to about 600 ohms. High-impedance mics usually have values at or above about 10,000 ohms.

High-impedance mics are seldom used professionally because they are much more subject to the pickup of hum and noise, and they can suffer from high-frequency loss if cables are more than a few feet long. In choosing microphones, a producer normally aims to match impedance between inputs and outputs. If the input to a mixer is low-impedance, then operators need to use mics having low-impedance outputs. Most preamplifiers and mixers used in audio production have low-impedance inputs or will readily accept low-impedance microphones.

Windscreens and pop filters are important tools for use with microphones. Windscreens are commonly incorporated in the mic itself, often in the form of a mesh diffuser placed in front of the diaphragm. External removable windscreens are widely employed as well, frequently made of a plastic foam material that blocks puffs of wind without affecting the transmission of sounds. Recordings made outside the studio are more likely to encounter wind noise problems, and a detailed discussion of windscreens can be found in Chapter 14, Field Recording. **Pop filters** function in the same way as windscreens but are intended for use on vocal microphones. These filters are made of a mesh stretched across a hoop. The hoop is held in place by a gooseneck mount that clamps to the microphone stand. The pop filter is placed between the performer's lips and the microphone, ordinarily at a distance of one to six inches.

Mounts are used to hold microphones in place and to attach them to a stand. Mounting can be accomplished in different ways, depending on an individual microphone's physical shape and design. Most often, some sort of mic clip is used to hold a microphone on a stand. These are usually plastic clamps that grasp the microphone securely and screw on the stand. If microphones are to be used in a high-sound-level environment it may be best to employ a shock-absorbing mount. This will prevent the transmission of vibrations through the floor to the stand and onward to the microphone pickup element.

Few producers take mic cables or connectors for granted because they are the source of more problems than the uninitiated might think. Experienced producers know that care in handling and storage of microphone cables will pay dividends in the long run. Cables should be loosely but neatly wrapped when not in use, then placed in cabinets or other protective enclosures. If properly wrapped, the cable bundle should unwrap itself without tangling when one end is tossed. Any suspect cable should be taken out of service immediately and repaired, otherwise it is likely to fail at the least convenient moment.

SPECIALIZED MICROPHONES

A number of specially designed microphones are available to meet the specific needs of certain production situations. Two will be mentioned here: the lavaliere and the boundary-effect microphones. One of the most well known, the lavaliere microphone is familiar even to nonproducers because it is one of the few mics seen by television viewers. These mics attach to a performer's clothing by means of a small clip, for instance, to a lapel or a blouse collar. To ensure that these mics are unobtrusive, they must be small and have a plain, flat, neutral finish. Some are as small as a pencil eraser. Older lavaliere mics sometimes had a loop attachment so they could be placed around the neck of a performer, like a necklace. It is from these that the mic gets its name, from the French word *lavallière,* which means "loosely tied bow."

Lavaliere mics are manufactured with special frequency-pickup characteristics, amounting to a boost of the upper frequencies and a cut at the extreme lower-frequency range. The odd frequency-response pattern of this type of microphone compensates for the loss of high frequencies when mics are placed off-axis from the speaker's mouth and for the tendency of a speaker's chest cavity to exaggerate low frequencies. Because of their peculiar frequency-response qualities, lavalieres—or "lavs" as they are often called—are not very useful in other applications.

When it is desirable to pick up sound from a distance within an enclosed space, boundary-effect microphones may be used. This sort of mic has the ability to pick up voices and instruments distant from the source without producing a severe off-mic sound. Boundary mics are manufactured with a "hemispherical" pickup pattern, meaning that it is like half of a sphere. Mics of this type have a flat profile, and are intended to be mounted on a flat surface. In fact, the flat surface is needed because it effectively becomes part of the mic itself. If a boundary mic is mounted on a table, sounds from all directions will be picked up equally—except below the surface of the table, of course. In other words, it has a pickup pattern of an omnidirectional mic cut in half. Boundary-effect mics are especially useful in picking up the voices of group members seated around a table, as in a meeting. These mics are also used from time to time in the studio for special purposes, such as miking a piano. The large size and variety of the piano's sound-producing parts make it difficult to capture adequately with a single mic, but the reach of boundary mics can encompass the entire instrument.

Stereo microphones represent a rather different sound pickup philosophy. Rather than capturing sounds directly from individual sources, stereo microphones attempt to reproduce sounds that a pair of human ears might hear at an appropriate location within the reach of multiple sources of sounds. Stereo microphones can be contained within a single housing, or may be devised by fixing two separate mics, usually cardioids aimed 90 degrees or more apart, on the same mounting hardware. Another kind of stereo microphone is the M-S or mid-side type. In this mic, there is a bidirectional mic pointed sideways coupled with a directional mic pointing forward. The outputs of the two mics are combined in phase relationships that simulate the stereo field.

If the quality of mics are good and their placement is correct, the result can be an impressive stereo effect. This is particularly useful when the number of sound sources make it difficult to obtain isolated pickup of individual sources; for example, in recording a symphonic orchestra. One important benefit of the stereo mic is that when located at a suitable distance, it will reproduce the **ambience** of the location, such as the sustained reverberation of a concert hall. Another obvious advantage of the stereo microphone is its simplicity. With no more than a stereo pair and a few additional microphones to fill gaps in the sound field, it is possible to make excellent-quality recordings of large-scale performances. A full discussion of stereo mic techniques is presented in Chapter 14, Field Recording.

MICROPHONE PLACEMENT AND OFF-AXIS COLORATION

The location of microphones in relation to sound sources has a large impact on the overall quality of audio recordings and broadcasts. Skilled producers are keenly aware of this and, to obtain desired results, apply their knowledge of how sound qualities of particular microphones are influenced by location and distance.

Microphone-placement decisions depend to a large extent on the type of microphone being used. There are significant advantages to omnidirectional mics, unless particular reasons exist for a directional microphone, because they usually produce a more natural sound. The reason for this is that the frequency response of most directional microphones varies radically depending on the the sound source's direction. A cardioid mic that has a reasonably flat frequency response at 0 degrees is likely to have poor highs and lows at 180 degrees. Hence, although this mic picks up far less sound from the rear, any that it does pick up will be reproduced inaccurately. This phenomenon is called **off-axis coloration** and is the inevitable result of designing a microphone that has a directional pickup pattern.

Despite the technical merits of omnidirectional or "omni" mics, they frequently cannot be used in a real-life recording situations. In fact, the majority of microphones used in actual recordings are cardioids. The reason is that most recording occurs indoors, and in enclosed spaces one is obliged to use pickup patterns to isolate sound sources so that they can be handled separately in mixing. If a dozen omnidirectional mics were used in a session, every mic would tend to pick up sounds from all sources in the studio and it would be impossible for levels of sources to be individually controlled.

When sound sources are at the rear of a directional mic, off-axis coloration can be a serious issue; however, coloration can also occur if sound reaching the mic has been reflected from walls. When sound bounces off a relatively flat, soft surface like a wall, higher frequencies are more apt to be absorbed than lower frequencies. This produces distortion of the frequency content of sounds and contributes to coloration of a mic's output. Mic placement thus must take into account not only the sources of sound that are desired, but the sources of sounds that should be rejected. Failure to do this will lead to pickup from unintended sources that reproduce poorly because they are too far away and off-axis from the microphone.

PROXIMITY EFFECT

When a mic is placed within inches of a sound source, a proximity effect may result. Sometimes referred to as *bass tip-up,* **proximity effect** produces an increasing boost of lower frequencies compared to mid-range and higher frequencies as the sound source moves closer to the mic. This phenomenon is most pronounced in ribbon mics, but is also present in any directional microphone, even cardioids. True omnidirectional mics are the only ones that do not demonstrate a pronounced proximity effect. To compensate for proximity effect, some microphones incorporate bass roll-off switches that permit a reduction in low-frequency response when placed close to sound sources.

Proximity effect can be used to advantage when it is desirable to boost low frequencies, as, for instance, in picking up a thin-sounding voice. Many vocalists like to perform close to mics to give their voices a greater impression of strength. But if a singer who has a deep bass voice is located too near to a mic, the proximity effect can produce a "boomy" quality, masking mid and high frequencies. In addition, a touch of proximity effect can be useful when picking up, say, a bass drum or cello, instruments that produce sounds rich in low-frequency content.

ACOUSTIC-PHASE RELATIONSHIPS

In the event that more than one microphone will pick up sounds, producers need to be conscious of acoustic-phase relationships. As noted in an earlier chapter, sound moves at about 1,100 feet per second, so sound picked up by mics on paths of different lengths will arrive at different points in their sound-wave-pressure cycles. Such differences in time relationships between two or more sound waves can cause acoustical-phase cancellation. If sound sources aren't placed correctly in relation to mics, sound waves will reach mics out of phase, possibly causing cancellation effects. For instance, imagine a performer placed off center between two microphones, spaced well apart. Because of differing path lengths, the same sound waves will reach one mic before the other, causing cancellation of their signals at certain frequencies when the two outputs are combined. When sounds are slightly out of phase, it can be difficult to perceive by ear. Listen for an irregular frequency response or a hollow sound. To avoid any possibility of this result, producers have traditionally followed the "3-to-1 rule." This guideline suggests that when pairs of mics can pick up a single sound source, the most distant mic must be no closer than 3 times the distance of the nearest microphone. If, for instance, the source is 6 inches from the nearest mic, the other mic should be no closer than 18 inches from the sound source. Some producers will extend this rule to 4-to-1 or 5-to-1, when sources produce high sound-pressure levels. Another trick to minimize the problem is to angle mics away from each other slightly but not off-axis.

DISTANT MIKING

Parabolic and hypercardioid mics are used for picking up sound at a distance from sound sources. Instead of the usual placement of a mic at distance of a few inches to 10 or more feet, these mics can produce sound of adequate—if not excellent—output level and quality much farther away, possibly up to distances of 100 feet. As noted previously, these mics usually impose sound-quality compromises: consequently, they are likely to be used only when closer placement is impossible. A good example of the correct use of this type of mic is in capturing field sounds at a football game. Even in a noisy stadium, mics like this can pick out the quarter-back's voice and the smacking sounds of football pads.

If equipment must be placed at a distance from sound sources and if the audio quality of a hypercardioid mic is not acceptable, it might be better to use a wireless mic. These are ordinary mics that can be placed close to the source, but designed to relay their output to distant equipment by radio transmission. Wireless microphones don't transmit very far, typically a few hundred feet at most, but this is usually enough to reach nearby production equipment. Television music productions frequently use wireless mics to pick up on-screen performers without unsightly cable runs. Handheld wireless mics also allow talent to move about without dragging mic cables along.

These rules and suggestions present producers with a multitude of options, some good, some not so good. The knowledge of which approach to take is founded not only on these well-established principles, but on experience. Seasoned producers possess hearing trained so that they can detect problem sounds, and they have learned from experimentation what works well for them in selecting and setting up mics.

CHAPTER FIVE

MIXERS

This chapter will broadly describe console operations and offer some general guidelines that can be applied to a variety of specific mixer applications. The **console**, **mixer**, or **desk** is typically the centerpiece of a recording studio, live-performance venue, or broadcast facility. Mixing consoles range in size and complexity from the simple, two-channel combiner that allows little more than level control to large automated computer-assisted consoles with hundreds of inputs and outputs and complex routing capabilities.

Even with the proliferation of digital audio workstations (DAWs), consoles still have value for the twenty-first-century recordist. A number of audio manufacturers have introduced the **control surface**, for use with DAWs, which emulate consoles in appearance and function.

PREAMPLIFICATION

At a minimum, mixing consoles do three things: amplify, mix, and route audio signals. As a signal enters the console, the first section it encounters is a **preamplifier** where low voltages presented by microphones are raised sufficiently high above the inherent noise floor to ensure good quality. In most traditional console designs, amplification occurs only within the input gain or preamplifier stage.

Whether used in recording, **live** sound, or broadcasting, correct setting of input gain levels is one of the most crucial steps in the use of a mixer. Proper gain adjustment makes certain that the console's full performance will be obtained, with optimum signal-to-noise ratio and dynamic range. If input signals are not given adequate preamplification, they will not be raised high enough above the audio chain's noise floor. Pieces of active electronic sound gear always generate a certain amount of circuit noise. That noise will be amplified along with intended signals as they move through succeeding sections of the sound system; each link in the signal chain will amplify noise along with desired signals. When signals are boosted above the noise in the preamplifier stage, listeners will hear less noise and more of the signals at the output—exactly what the producer wants. If the input gain levels are too low and the operator attempts to compensate by pushing the faders up to boost the console's output, it will shrink the signal-to-noise ratio—exactly what the producer does not want. In Figure 5.1, the upper illustration shows the result of raising only the output level—noise is increased along with the signal. The bottom illustration shows that increasing the input gain raises the signal level but does not raise the noise level, resulting in a greater signal-to-noise ratio.

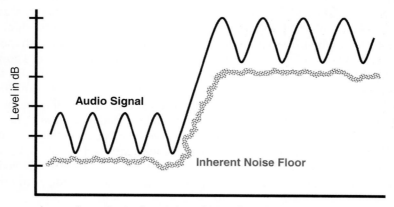

Increasing output volume also raises noise output.

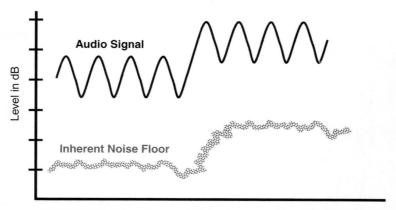

Raising input gain in the preamplifier increases the difference between noise and signal.

FIGURE 5.1   Top: Setting gain levels too low will not sufficiently raise the signal above the noise floor. Bottom: Proper input gain settings improve the signal-to-noise ratio, and allow the output signal to be raised above the equipment's inherent noise.

PEAK INDICATORS

Most recording and live sound consoles have a simple rotary control that is used to adjust the input gain. Figure 5.2 shows an input gain pot on a recording console. In broadcast consoles, gain adjustments are normally set internally, and are not user accessible. The input gain control regulates the gain of the channel pre-amplifier. To aid engineers in achieving optimum gain settings, most consoles also have clip or peak indicators in the form of light-emitting diodes, or **LEDs**. Some consoles incorporate two LEDs on each channel; often one is green and the other is red. Other consoles have a single red LED. The green LED indicates that a signal is present, and the red LED indicates that the signal has exceeded the gain amplifier's clipping threshold. Clipping produces distortion of the audio signal.

The term **clipping** derives from the fact that an audio wave shape will have its peaks "clipped" off or flattened if the signal exceeds the channel's amplification capacity. If clipping occurs to a sine wave signal, as an example, the nicely rounded sinusoidal shape of the wave will be squared off. The resulting squared wave has an unpleasantly harsh and distorted sound. In the illustration in Figure 5.3, horizontal lines indicate an audio channel's capacity. As the signal is pushed beyond that limit, it can be seen that the waveform becomes flattened or clipped.

As critical as it is, correctly adjusting input gain is a simple operation. With the signal present at the console's input, the gain level control is raised until the red clip indicator blinks *occasionally*. Its blinking should coincide with audio-signal

FIGURE 5.2   The gain trim pot on a recording console's input channel is shown here, along with input level indicators. If the red peak indicator glows steadily, the gain is too high and distortion will result.

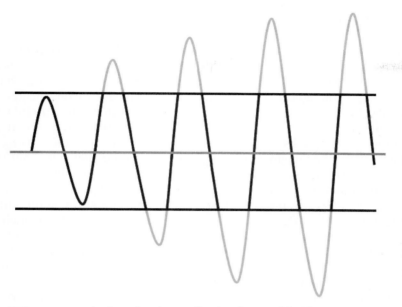

FIGURE 5.3   As the gain of an audio signal is amplified beyond a channel's capacity, the waveform will be clipped.

peaks. If the clip LED is not flashing, the signal is probably too low to achieve full signal-to-noise performance and thus has more noise than necessary. If, however, the clip LED is steadily illuminated or remains lighted for extended periods of time, the input level is set too high, and the signal will show signs of clipping and distortion. An example of a clipped signal can be heard on the **CD-ROM** accompanying this text.

Console input modules have clip thresholds that vary from one model to another, so some trial and error will be needed to discover a mixer's exact clipping point. To do this, signals should be intentionally forced into clipping until distortion can be clearly heard. Then, reduce gain just enough so that signals sound normal once more and note how frequently the LED flashes. Again, this point will vary somewhat, depending on the mixer's design. In subsequent uses of the mixer, gain levels can be set properly by adjusting input levels so that the same LED blink rate is obtained. Great care should be taken to ensure that this is done accurately.

LED indicators are handy visual referents, especially on a mixer familiar to an engineer, but a signal's sound should be the final determinant of proper settings. Signals must be clean and free of distortion, and system noise should be unobtrusive. Gain adjustments may be needed during a performance or recording session, as speakers and performers often increase their sound levels gradually as time passes. Finally, this caution should be kept in mind: errors in level adjustments will introduce noise or distortion that *cannot* be removed at a later stage.

FADERS AND POTENTIOMETERS

A console's mixing function is accomplished through the use of a **potentiometer**. Most contemporary consoles employ **linear** potentiometers that are called **faders** or sliders. Some smaller mixers, and especially powered mixers that have power amplifiers built in, do still employ rotary potentiometers. Mixers incorporating linear faders have the advantage of presenting an operator with a useful visual indication of relative levels among channels. Linear faders are available in a range of throw or travel lengths, from short (75 mm) to long (120 mm). Longer throws make it easier for an engineer to achieve a slow smooth fade or to make small level adjustments. High-quality long-throw faders are considerably more expensive than short-throw types that are more common on inexpensive mixers.

Some faders are motorized to facilitate console automation. Often called **flying faders**, motorized types are linear faders activated by a small electrical motor. A servomotor is a device that has its movement monitored and controlled by a feedback circuit. Servomotors are commonly used for remote controls (such as in radio-controlled models, where a motor moves the wing flap or steering wheel). A signal is sent to the motor instructing it to make a specified number of revolutions to position itself precisely along a fader's throw. The motor's feedback loop monitors the number of revolutions to ensure accuracy. Penny & Giles Corporation developed the first designs of moving faders as an adaptation of a standard linear fader. Essentially, P & G adapted the 3200 series linear fader to a small servomotor connected with a cord to drive automated movement of the fader. Subsequent motorized fader designs have been in some way derivative of the original motor-and-cord design. In console automation, fader positions are calibrated and stored as a count of motor revolutions. For instance, moving a fader from off to half gain might require the motor to revolve twelve full clockwise revolutions. This information can be stored so that the fader can return to its half gain position by activating the motor and instructing it to make twelve full clockwise revolutions from off.

It is best to think of a fader as a gain reduction device. When the fader is pushed fully up, it will not reduce signal level at all. But as the fader is pulled down, it will successively reduce its output signal more and more. When the fader is pulled down completely, signals will be totally attenuated so that audio can no longer be heard. Therefore, while input modules boost the signals to their optimum levels for dynamic range and signal-to-noise ratio, faders reduce that volume to a level suitable for listening or recording. Legends on many faders indicate 0 when the fader is "full up," meaning that no attenuation is supplied in that position. At the lowest settings legends may be labeled ∞, indicating that attenuation is infinite or total.

Input channels on a console may be designed to accept either a stereo or monaural signal. In the case of a stereo input signal, such as that from a CD player, tape machine, or computer sound card, a single control mechanism attenuates both the left and right stereo signals simultaneously through twin faders. The relative balance between the left and right channels is adjusted by the pan control, discussed later in this section. Broadcast consoles normally contain many more stereo channels than monaural channels, but the reverse it true for recording and live sound consoles.

LEVEL INDICATORS

Mixing is normally aided by the use of output signal level indicators. Indicator types commonly found on contemporary consoles include the traditional analog meter with moving needle, light-emitting diode (**LED**), liquid crystal display (**LCD**), fluorescent meters, or the less common **plasma** gas **display.** Operators should become familiar with the characteristics of each display type, especially the speed with which each reacts to changes in level. Not every **averaging meter** is designed to respond to **transients**, or short spikes in levels. The ability of a needle-type meter to indicate transients is determined by its mechanical characteristics. An averaging meter reacts slowly—on the order of 300 milliseconds—to indicate an average of signal levels. Averaging meters may incorporate peak indicators to warn of possible distortion from excessively high transient levels. The **peak program meter (PPM)** responds very quickly—approximately 5 to 10 milliseconds—to display instantaneous transient peaks. LED, fluorescent, and plasma indicators sometimes have their functions user-assigned so that operators can choose to display either average or peak levels. Some meters can even indicate both simultaneously. Figure 5.4 shows a variety of meter types, including VU, dB, LED, and fluorescent.

FIGURE 5.4   A few common types of level indicators are shown here, clockwise from left: segmented LED meter, peak meter, plasma meter, segmented LED meter; Center: standard VU meter.

Meter scales can be marked in several different ways. Scales commonly found on level indicators include the volume unit or **VU meter**, the **percent of modulation** meter (frequently used on broadcast consoles), and the decibel or dB meter. Sometimes level indicators may only display digits 0 through 10, and have no real unit of measurement reference.

The VU meter was collaboratively developed in the late 1930s by Bell Laboratories and U.S. broadcasters. It is an averaging meter. A VU meter approximates the **root mean square (RMS)** value of an audio wave's electrical signal. Root mean square is a statistical measurement of the effective power of complex waveforms of the sort encountered in audio. RMS allows AC signals to be computed as equivalent to DC values. The VU meter, like most level indicators, was designed to respond to changes in loudness in much the same way as the human ear. The VU meter's readings represent a smoothed value of variations in the audio signal's level. Most VU meters place a VU scale above a percent of modulation scale. Some meters' legends feature a dB scale, but operators should not regard these scales as necessarily accurate. Because the dB is a relative measure of loudness, indicators must be carefully calibrated to the system's acoustic output in order to obtain accurate readings.

Calibration is usually carried out by studio maintenance engineers, and though not difficult, it is time-consuming. One simple method of calibration uses a signal of constant level input to the console, possibly a 1,000 Hz sine wave. Consoles often have their own built-in oscillators for calibration purposes. The output fader is raised until the console's output meter reads 0 VU, or 100 percent modulation. After this, input levels of every recording device in the studio are adjusted so that each of their level indicators also read 0 VU. The operator is thus assured that when the console indicates a specified signal level at the mixer, each recorder will receive exactly the same level. When some devices in the signal chain use averaging meters and others use PPM meters, compensation must be made to ensure that they provide consistent indications. A rough rule of thumb commonly used by engineers is to calibrate PPM meters −8 dB below averaging meters. Following this guideline, if the input level to an averaging meter device is 0 VU, the input to the PPM meter should be set 8 dB lower on its scale.

HEADROOM

To further ensure there is no distortion in the recorded signal, the system's **headroom** must be determined. Headroom is designed into an audio system to provide a margin of safety in case signals exceed 0 VU. It is measured in the total dB a signal can exceed 0 VU before clipping begins. To measure a system's headroom, raise a steady signal above the 0 VU mark on the meters until distortion can be heard. Note the point at which distortion occurs, then compute headroom as the difference between this level and 0 VU level. Experienced engineers do not rely only on indicators to set levels also but also on their ears. Meters are an aid, but, as has been stressed previously, the sonic integrity of a signal is the final determinant of proper levels. In systems that are properly calibrated, signals having frequent transients may not sound distorted even though meters appear to be "buried in the red."

SIGNAL ROUTING

A console's third and final function is routing, or directing the signal out of the console and on to its next destination. Modern recording and live sound reinforcement consoles include several pathways or **buses** for the signal to take as it leaves the mixer. (Readers may find that some manufacturers and authors use the spelling "bus," while others use "buss.") For example, all consoles will have some sort of main output. This signal is typically routed to an amplifier or a recording device. In many modern consoles, the main output may optionally be referred to as a **stereo** output or a **two-track** output, since most consoles and other audio devices are designed to operate in a stereo mode. With the individual channel levels adjusted relative to one another by the channel faders, the main output is usually adjustable by a master fader, or perhaps a pair of faders. Master faders provide attenuation for the overall mixed signal. The console depicted in Figure 5.5 incorporates a single stereo fader (on the far right) that adjusts the overall mix.

FIGURE 5.5   An audio console providing stereo output, controlled by a single red fader on the far right.

SIGNAL SENDS

Most consoles also allow signals of individual channels to be routed to other destinations. Both live sound and recording consoles normally incorporate alternative buses or **sends** that may be connected to external effects equipment, foldback monitors, headphones, or other devices. Nomenclatures used for these buses include **auxiliary,** cue, **monitor**, or **effects**. The number of these alternative buses varies greatly depending upon complexity of the mixer and its intended usage. Figure 5.6 illustrates the wide selection of inputs and outputs that may be routed through a recording studio console by means of different buses.

The signal path for these auxiliary buses is often selectable between pre and post by a switch in the send section. This terminology indicates whether the signal sent to the bus is **pre-fader** or **post-fader**, and **pre-equalizer** or **post-equalizer**. In the *pre* position, signals routed to the auxiliary output bus are

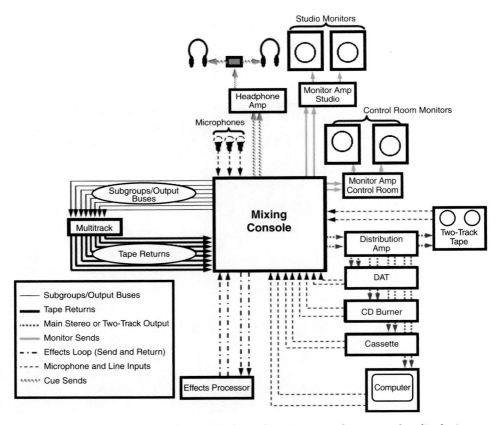

FIGURE 5.6   This typical configuration shows how inputs and outputs of audio devices are routed through a mixer in a production studio.

unaffected by either the equalizer section or channel faders. In the *post* position, signals sent to the auxiliary output bus will first pass through the equalizer and fader. Figure 5.7 illustrates the difference in signal paths between pre-fader and post-fader sends. Refer also to Chapter 12, Live Sound Reinforcement, for additional discussion of this feature.

Consoles that are primarily designed for multitrack recording usage provide a series of output buses or **subgroups** that can be connected to the track inputs of a multitrack recording device. Signals of individual channels may be assigned to any of the subgroups by a **bus assign** or group assign switch. Some consoles may also include a series of subgroup or output faders that allow combined signals to be adjusted as they leave the console. Consoles incorporating this design are sometimes called **split consoles**, or less frequently **I/O** (for input/output). The split designation refers to the mixing surface being divided into two halves. By convention, the left side is devoted to inputs and the right side is arranged for output groups. This approach is particularly useful if more than one input channel is assigned to a single bus, as would occur when an engineer has a limited number of

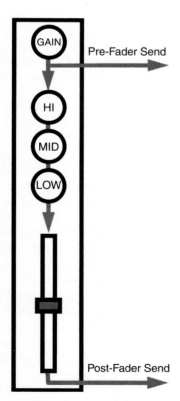

FIGURE 5.7   Pre-fader sends bypass the fader, while post-fader auxiliary sends allow the fade to control levels.

FIGURE 5.8   The outputs of individual channels may be routed to selected subgroups, where those signals are summed and levels adjusted in the subgroup section of a multitrack mixer.

tracks. The subgroup fader allows an operator to adjust overall track levels. The subgroup faders on a mixer with four output buses or subgroups are highlighted in the photograph shown in Figure 5.8.

SPLIT AND INLINE MIXERS

The main difference between split and **inline consoles** (occasionally called "British" mixers) is the absence of group faders on the latter. In these consoles, each input channel is **normalled** (normally connected) to its associated recorder track output. In other words, microphone one is normalled to track one, microphone two to track two, and so on. Each input channel still has bus assignment switches to route signals to alternative output buses. Levels of the output buses may be adjustable, but a rotary trim knob on input channels is used to set individual levels rather than by means of a separate fader section. For instance, microphones connected to channels one and two may be assigned to track four. The rotary trim knob on channel four will set the level being sent to track four.

Both consoles types have certain advantages. Split consoles are possibly a bit more intuitive for novice operators. The layout of the console surface provides a simpler visual metaphor for signal flow. Inline consoles are more compact, and offer more channels in a smaller frame.

Audio professionals refer to consoles by their input channel and output bus configurations. For example, a "twenty-four-by-eight" (24 × 8) console has twenty-four input channels and eight output buses. Similarly, an eight-by-two

(8 × 2) mixer has eight inputs and two outputs. Manufacturers frequently incorporate the number of input channels and output buses as part of the model designations of their mixing consoles.

Mixers usually provide a set of output buses labeled cue, monitor, or foldback for the purpose of sending signals to headphones in studio usage, or to stage monitors in live performance situations. Most consoles only provide one or two sends like this, but some might have a dozen. The most elaborate recording consoles may even provide stereo cue sends, making it possible for headphones used by studio performers to receive true stereo signals.

Other output buses labeled effects send are intended to be connected to **outboard** effects devices such as reverb, delay, or multi-effects processors. Effects sends are customarily selectable between pre- and post-equalizer and fader, just like cue sends are. Console manufacturers sometimes merely label outputs as auxiliary, giving no hint of the possible uses of their externally routed signals.

CONTROL ROOM MONITORING

Consoles that are solely designed for recording studio uses sometimes designate one set of outputs as **control room monitor** sends. These can merely be individual sends on each input channel, but some consoles set aside an entire mixer section for the control room mix. In either case, the control room sends are intended to be used during recording sessions to set up rough mixes for control room monitoring purposes. This application is described in more detail in Chapter 15, Multitrack Recording.

MUTE AND SOLO

Most consoles incorporate **mute** and **solo** features that enable engineers to isolate the signals of individual channels. As the reader might guess, the mute switch is intended to silence a channel. A solo switch, on the other hand, silences all channels except the one on which it is activated. This permits a selected channel to be heard alone in the control room monitor without the distraction of sounds passing through other channels.

There is one extremely important difference in the operation of these functions on a recording console: the mute function affects signals going to the tape, while solo only affects signals being sent to the control room monitors. As a consequence, when engineers solo a channel in a recording studio, signals being captured on the recording device are unaffected; only what they hear is influenced. In contrast, muted channels do not go anywhere; the mute function blocks signals from being either recorded or heard.

Sophisticated consoles may also feature programmable or grouped mutes. This function allows an engineer to mute a set of preselected channels, a feature that is useful for critical monitoring of recorded passages in which a solo instrument can be isolated while the other performers are silenced. For instance, if in a recording session an acoustic guitar plays a lead-in before the remainder of a musical group joins in, muting all but the acoustic guitar channel will prevent background noises produced by amplifiers and mechanical noises of the drum kit from being recorded. The solo function cannot address this problem for, as just noted, it affects only sounds operators hear in monitor speakers.

PAN POTENTIOMETERS

A **pan** potentiometer, short for panoramic, is provided on mixer channels for adjustment of signal levels sent to left and right inputs of the main output or stereo bus. Pan pots may also be used to adjust output levels routed to subgroups. An arrangement commonly seen on consoles is a single assign button for sets of paired subgroup buses. These are often paired as subgroup buses one and two, buses three and four, five and six, and so on. Pan pots adjust the relative balance in level between selected pairs—for example, bus one and bus two—just as it might balance signal levels in the left and right channels of a stereo mix. Figure 5.9 shows the pan section of a console in which the left and right channels are labeled 1 and 2, respectively, and eight subgroup buses are designated A through H.

FIGURE 5.9   The pan pot and bus assign section is used to route the output of a channel to the left and right stereo subgroups or to individual output buses.

Most recording and live sound consoles also incorporate a selectable pad for each channel. A **pad** is a gain reduction circuit that can be switched on when operators wish to cut the level of input signals by a fixed amount. Pads are more fully described in Chapter 9, Audio Processors and Processing. In addition to pads, consoles may also include phase reversal switches, which can be used to correct cabling with reversed polarities.

CHANNELIZED MIXERS

The features described above are usually integrated by organizing the console into **channel strips**. Channel strips include ones designated as input channels, subgroup bus channels, master output channels, and the like. In some **modular consoles**, channel strips may be added, removed, or combined to fit the needs of particular production applications. Modular consoles offer users the freedom to remove and replace individual channel strip modules by simply plugging in new ones. In nonmodular mixers, channel strips are permanently fixed in place, although they are still arranged as discrete strips organized by function. In Figure 5.10, two consoles are pictured, both having nonremovable fixed channel strips. At the top is a small analog mixer and at the bottom is a digital console with fixed channels, but in which routing and equalization functions are adjusted in assignable sections. In Figure 5.11, a modular console with a single channel strip removed is shown.

The layout of channel strips seldom differs from console to console. Each input strip will be laid out beginning with the input gain control at the top, followed by the equalization section a bit lower, then the bus routing assigns and pan controls next, and finally the fader at the bottom. Uniformity in console layout makes it easy for an engineer to operate an unfamiliar mixer, so it is often said that if "you learn to operate one console, you can truly operate any console."

MIXER AUTOMATION

Some mixers include **automation** facilities to ensure consistency in complicated mixes. These systems work by continuously recording levels as a set of fader settings. Console automation can be categorized as either *scene* automation or *dynamic* automation. Incorporating scene (or snapshot) automation in a mixer is straightforward when servomotor mechanisms are used. As the term *snapshot* suggests, a record is made of level settings as motor positions at a chosen moment. When fader positions are recalled, they act as instructions for the fader motors, telling each one how many revolutions are required to move its fader to the recorded position. Dynamic automation updates fader position continuously—or more accurately, at regular, short intervals. Because much of audio, video, and film synchronization is referenced to a standard coding scheme established by the Society of Motion Picture and Television Engineers (**SMPTE**), many systems base

FIGURE 5.10 Two mixers constructed with fixed channel strips.

update timing on SMPTE time code. In effect, time code acts as a clock that initiates an update on each tick in order to create an event **timeline**. SMPTE time code is based on a frame rate of approximately thirty frames per second, or fps (actually 29.97 fps). Use of this standard ensures that motor positions are stored and read thirty times each second, frequent enough for any likely application. For a full discussion of time code see Chapter 16, Sound for Pictures. Dynamic automated mixing requires that the console be synchronized to a recording device. The console's internal memory processor stores fader, equalization, auxiliary level,

FIGURE 5.11   A mixer utilizing modular removable channel strips, with one channel strip removed.

and perhaps even other settings following time code received from the recorder. The clock that triggers recording of each snapshot is synchronized with time code sent by the recorder. On playback, the console reads the recorded time code and adjusts mixer controls to the stored values thirty times each second.

More and more consoles are using electronic instead of mechanical controls for automation purposes. These mixers utilize a **voltage controlled amplifier (VCA)** that operates under control of software algorithms, rather than traditional potentiometers and faders. In these mixers, changes in level and equalization are performed during an automated mix down, but no controls physically move. The key advantage of channel faders having servomotors is that they provide visual cues of the relative levels and to changes that occur, primarily to assist operators. In nonmechanical automated mixers, moving faders can be simulated in the software of computer-based recording systems via on-screen displays.

DIGITAL CONSOLES

Another recent innovation in mixing console design is the **digital console**. These mixers use microphone preamps to raise low input voltage levels prior to conversion into digital data, but all other functions are carried out entirely in the digital domain. Many contemporary professional digital consoles incorporate analog-to-digital converters that have sampling rates of at least 96 kHz and bit rates of 32 bits. Once the analog signal is converted to a digital bit stream, nothing in the signal path will add noise or distort the audio waveform. Changes to levels, equalization, stereo panning, and countless additional audio effects can be internally performed using a mixer's preprogrammed digital algorithms.

An important advantage of digital consoles is that signals received from or sent to digital devices such as DAT players or DAWs need no analog conversion. Audio in the form of binary data stays in digital form as it travels through the console to its output, usually to a workstation or digital recording device. Since digital mixers do not use hardwired buses but employ software-based virtual circuitry, they can be configured in practically any imaginable arrangement. Any and all inputs can be routed to any and all outputs through any selected level or processing control. For example, additional effects sends can be easily activated as either pre-equalizer (EQ) or post-EQ virtual devices. Digital consoles can incorporate all sorts of onboard effects, even including reverberation, delay, and chorusing. This flexibility does come at a price—digital consoles must employ powerful digital processors in order to handle the complex programming and massive flow of data.

The enhanced routing capabilities available in digital consoles can be a major advantage in applications such as mixing for surround sound. The six discrete submixes or channels required for Dolby 5.1 can be easily created in a virtual console and assigned to defined outputs. This is such a common usage that many digital consoles now include preset assignments for 5.1 mixing. Most digital consoles have an assignable control device such as a wheel or joystick that can be designated as a pan control to balance signals among the six surround channels. Surround mixing is discussed further in Chapter 16, Sound for Pictures.

BROADCAST CONSOLES

Broadcast consoles, although adhering to the general design principles outlined throughout this chapter, have certain distinguishing features. As mentioned previously, broadcast mixers usually do not have user-adjustable input-gain levels. Broadcast consoles have no need for a large number of output buses such as those required for multitrack recording or live-sound consoles. Broadcast consoles employ much-simplified control surfaces that eliminate most of the switches and pots used in recording and live-sound mixing where greater flexibility is needed. The photograph in Figure 5.12 shows an example of the console surface in a radio broadcast control room. Routing assignments and level controls are **hardwired**, preconfigured by the broadcast engineer, not the operator. Figure 5.13 shows the signal flow diagram for a small radio broadcast console.

FIGURE 5.12   A broadcast mixing console has fewer user adjustments because input gain levels and equalization are adjusted internally.

In broadcasting, the main output of a console is sent to a transmitter chain via a distribution amplifier. The output signal may also be routed to recording devices that capture programs for production purposes or for recording an **air check**. A special feature of broadcast consoles is the **cue** function. Cue here refers to a signal that is routed to an amplifier and speaker purely for the purpose of previewing program material prior to broadcast. On channels equipped with this function,

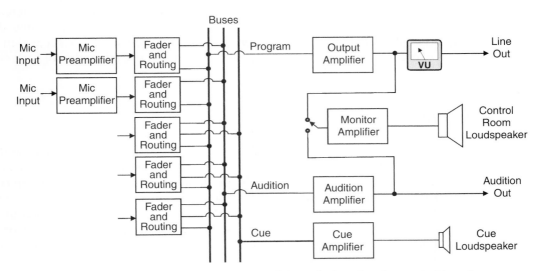

FIGURE 5.13   All audio signals are routed through the mixer in a typical broadcast station.

completely pulling down or rotating its fader to off will send the channel's output to the console's cue bus. In this position, a **detent** or click stop simultaneously switches off the channel and routes the signal to the cue amplifier and cue monitor system.

Broadcast consoles also provide an on-off switch for each channel. The switch frequently has multiple functions. Most often it may activate the channel while simultaneously activating whatever device supplies its signal. For instance, when an operator applies a switch to activate a compact disk player channel, the channel output will be connected to the main output bus and at the same instant the CD player will be started. On channels assigned to control-room microphones, the switch will activate the channel while muting the control room monitor speakers to prevent audio feedback.

Even though broadcast-console operators have no ability to control channel input gain, managing output levels is even more important in broadcasting than in live sound or recording. Overdriving console outputs in live sound productions might result in feedback, and overdriving output of a recording console might cause distortion, but overdriving the output of a broadcast console will almost certainly result in overmodulation and serious distortion of broadcast signals. For this reason, broadcast operators must be intensely aware of console output levels and observe level-indicator readings closely.

AUTOMATIC CONSOLES

A few other somewhat specialized console designs can be mentioned briefly here. An automatic—*not* automated— mixer is really a console with a gated compressor-limiter or compander incorporated in the audio chain. Automatic mixers are used almost exclusively in large convention and meeting facilities, where several microphones are connected to a sound system. This mixer is designed for unattended operation. In use, the automatic mixer opens a channel's gate only when the input signal exceeds a preset threshold and the compressor-limiter keeps signals from exceeding a preset maximum level. For a complete explanation of such systems see Chapter 12, Live Sound Reinforcement.

FIELD MIXERS

Portable field mixers are constructed to suit the particular needs of location sound recordists. Applications for the field mixer include electronic-video field production, location film production, electronic newsgathering, and **bioacoustics** recording, among others. These are discussed in Chapter 14, Field Recording. Field mixers normally offer only a few input channels, usually six or less. Requirements of field mixers include rugged construction, compact design for ease of transport, and the provision of long-life internal battery power. A means of easily carrying and supporting the mixer may be provided as well, perhaps a neck strap that leaves the recordist's hands free to operate controls. Most field mixers incor-

FIGURE 5.14   A portable mixer intended for use in field production can be powered by batteries in the field.

porate rotary potentiometers, rather than linear faders. Rotary controls are more compact and are not as easily bumped off their settings as are linear faders. Controls on these mixers may be recessed and heavily dampened as a further measure against accidental bumps. The portable mixer in the photograph in Figure 5.14 operates on two ordinary 9-volt batteries to facilitate field use.

VIRTUAL MIXERS

Computer software interfaces or virtual mixers, or control surfaces as they are also known, offer engineers a familiar visual and tactile mechanism for interacting with recording software applications on digital audio workstations (DAW). Control surfaces are used to replace computer mice and keyboards with devices more familiar to audio operators. The controllers are made to have the appearance of small mixers, including faders, knobs, push buttons, and connectors, but the interface is merely used to control parameters within a workstation's software. In some units the interface provides analog-to-digital converters for inputting a digital data stream to the workstation. Because the controller actually relies on and directs software instructions, users may be able to assign functions to some controls. A single rotary control called a data wheel may be included on the interface to set different parameters of the workstation. Most interfaces include shuttle controls that allow the operator to manipulate play, rewind, fast-forward, and record functions just as if the recorder were using tape. Controllers commonly connect to workstations through an **IEEE 1394 FireWire** or **USB** connection.

CHAPTER SIX

RECORDING, STORING, AND PLAYBACK OF SOUND

In little more than a century, techniques for audio recording have moved beyond their primitive origins to sophisticated methods of translating sound waves into audio formats capable of astonishing realism and clarity. Key points of this evolution include a shift from purely mechanical to electromechanical methods of recording, to magnetic recording, and, finally, to a host of non-analog, digital recording techniques.

PHONOGRAPH RECORDING

The earliest methods for sound recording and storage were mechanical in nature. Most historians attribute the development of mechanical sound recording to Thomas Edison. Edison's "talking machine," introduced in about 1877, was called the **phonograph** or, literally, the "voice writer." In its original incarnation, the phonograph used a foil-wrapped cylinder as a recording medium. The performer would speak or sing very loudly into a funnel-shaped horn. At the small end of the horn, a stylus was rigged to move in concert with the sound vibrations traveling through the horn. The stylus created an embossed groove in the foil. During playback, grooves in the cylinder caused the phonograph stylus to vibrate exactly as the stylus did when recording the cylinder, and the vibrations were focused by the horn so that sounds would be audible to nearby listeners. This kind of recording mechanism is shown in Figure 6.1.

Cylinder recordings gave way to something more recognizable to modern music lovers—a disk—which Emile Berliner developed as the *gramophone* around 1888. Berliner's machine employed the same horn-stylus arrangement to cut a groove into a ten-inch flat disk made first of glass, then zinc, and finally a plastic material. The gramophone disk rotated at 78 revolutions per minute or **rpms**. (Figure 6.2 shows Berliner and his gramophone.) Berliner eventually sold the rights to his machine to the Victor Talking Machine Company, which would later become known as RCA Victor.

The technique of cutting a spiral groove onto the surface of a disk has remained relatively unchanged over the years. Significant innovations in the tech-

FIGURE 6.1   The early Edison cylinder phonograph, or graphophone, was a purely mechanical device capable of both recording and playback. It was originally intended to be used as a dictation recorder.

nology have included the introduction of the twelve-inch $33\frac{1}{3}$-rpm long play (**LP**) **microgroove** record that extended playing time of disks and enabled a single one to contain an entire album of as much as one hour of recordings. Columbia Records developed the Microgroove LP in 1948. RCA Victor introduced the seven-inch 45-rpm disk in 1949, largely as a promotional tool aimed at the radio and jukebox markets. This was often called a "single" because a single musical piece of up to about five minutes could be placed on one side of the disk. Other variations on the disk have included the ten-inch extended-play single, and the twelve-inch single. But still, all these types of records are essentially a spiral groove cut into a plastic or plastic-like disk by a modulated stylus.

Even though vinyl disk recordings have greatly faded in importance as a consumer format, they are still popular with a small market segment. On playback, disk recordings make use of the kinetic energy provided by **stylus** movements that follow the tracings in the plastic disk. In stereo recordings, the left channel is cut into the side of the groove nearest the center of the disk while the right channel is nearest the outside of the disk. The width of the groove's excursions is dependent on the instantaneous signal amplitude—the higher the signal peaks, the wider the swings of the groove.

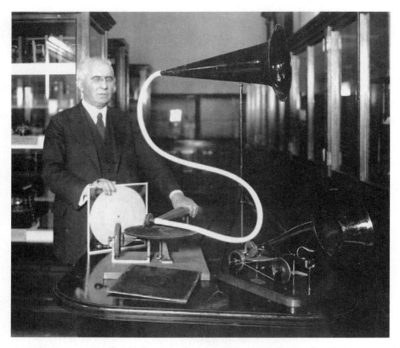

FIGURE 6.2   Emile Berliner is shown with an early mechanical gramophone, a recorder/reproducer that made recordings on flat disks, rather than the cylinders used in Edison's early machines.

To properly play a recording, the disk must be rotated by the turntable at a steady speed of 33.33 or 45 (or perhaps 78) revolutions per minute. If the speed fluctuates, the pitch of sounds will waver noticeably, something that is highly objectionable to the ear. The pickup cartridge produces an electrical signal as the stylus traces the disk's groove. Figure 6.3 shows the construction of a simple monophonic **phono cartridge**. The stylus is attached to a tiny powerful magnet that is placed within a coil fixed to the pickup. The stylus traces the disk as it rotates, and this forces the magnet to move as it follows tracings of the groove. Movements of the magnet yield a voltage output at the coil that matches the groove's twists and turns. This voltage reproduces the original recording and after amplification can be heard by listeners. Stereo pickups are similar to the one shown in Figure 6.3, but they have two coil and magnet assemblies, one each for the left and right channels.

One serious problem in vinyl recordings is hiss, the scratchy sound produced as the stylus slides along the groove. To reduce this noise, disk recording makes use of a system of preemphasis of high frequencies on recording and a parallel deemphasis on playback. The response curves used for preemphasis/deemphasis

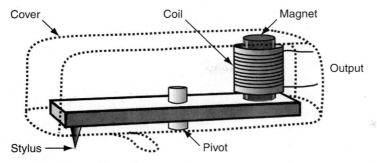

FIGURE 6.3   The stylus in this monophonic phonograph pickup cartridge vibrates as it rides along the grooves of a record. Small movements in the magnet attached to the pivot arm produce an electrical current in the coil.

are known as the RIAA response curves. Use of this system cuts disk noise by as much as 18 dB.

MAGNETIC TAPE RECORDING

The heart of much of modern sound storage is magnetic recording. Oberlin Smith may have first demonstrated the principles of magnetic recording in 1888. Smith used a telephone to convert sound into electrical current, and then used a string coated with iron filings passing through the field of an electromagnet, to store a magnetic of the audio signal. No working model of Smith's device has survived to the present.

   Most recording historians credit Valdemar Poulsen with the development of magnetic recording. Poulsen patented a machine that he called the Telegraphone, which used steel wire wrapped around a drum to store magnetic patterns. Poulsen's, and all subsequent magnetic recording devices, are based on the principle that electrical current passing through an electromagnet will create a magnetic field that can be impressed on **permeable** material (one that can be magnetized) passing through the field. The strength of the magnetic field is proportional to the audio current, and therefore the field retained by the permeable material is a replica of the audio current. Steel wire recorders were distributed commercially for about a decade or so but were soon replaced by **magnetic tape** recorders because tape was much more durable and easier to use. The first patent for a paper tape coated with magnetic powder was issued to Fritz Pfleumer in the 1920s, but tape was not widely used until the late 1930s. In fact, until the mid-1940s, wire recorders and tape recorders were both in use, but by 1947 tape had become the exclusive recording medium, with companies like Minnesota Mining and Manufacturing (3M) producing tapes, and Ampex, Magnecord, and others producing recorders.

Magnetic tape recorder technology has not changed much over the years. In recent decades there have been refinements in tape formulations, improvements in the materials from which heads are manufactured, and advances in the mechanical construction of machines—especially mechanisms that move the tape. However, the basic principles of recording on magnetic materials have remained, whether speaking of a cassette recorder, digital audio tape recorder, or even the hard disk in a computer or digital audio workstation.

During the recording process, an alternating electrical current—an analog audio signal—is applied to the **recording head**. The head is constructed from a coil wrapped around a permeable material, creating an electromagnet. A magnetic field is created and focused at the **head gap**. As the applied current alternates, the magnetic polarity of the field alternates, changing the polarities of the magnetic particles on the recording tape. During playback, the **playback head**, which is constructed much like a recording head, detects fluctuations in the magnetic properties of the tape particles, and from these changes induces an alternating current flow in the coil. This current will be a faithful copy of the current that flowed in the recording head. The underlying principle is the relationship between magnets and coils, wherein coils within a fluctuating magnetic field have induced within them a current flow, and current flowing through a coil will create a fluctuating magnetic field. How this works can be seen in the drawing of Figure 6.4.

Head size and gap vary greatly according to the recording medium. Analog multitrack tape machines require large heads, sometimes more than two inches in width, while the head on a fixed disk in a computer is the size of a pinhead.

Early recording tapes were made from paper, with a substance like powdered rust affixed to one side of the strips of paper. Today's tape recording makes use of tape made of a layer of an easily magnetized material (usually a powdered ferrous oxide compound) bonded to a long strip of tough plastic—usually acetate, Mylar, polyester, or, less frequently, polyvinyl chloride (PVC). These materials comprise the backing, onto which the oxide is "painted," and held in place with an adhesive **binder**. Because the recording surface is made up of very tiny bits of magnetic material, each small particle can act as an independent magnet. The size of mag-

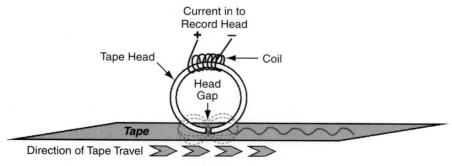

FIGURE 6.4   A tape head fixes the magnetic polarity of oxide particles on the magnetic tape as it moves past the head.

netic particles must be smaller than the width of the recording head gap in order to accurately retain magnetism they receive from the head. Audio tape is manufactured in widths of quarter-inch, half-inch, one inch, and two inches, but the most frequently used has been quarter-inch.

A major complication in the recording process is that magnetic tape has nonlinear magnetic transfer characteristics. This means that the relationship between the applied magnetic field and the magnetic field retained on the tape is not exactly proportional, especially around zero magnetism. If recordings were made without taking this into account, serious distortion would result. The problem is due to the tendency for the magnetic field in permeable materials not to fully collapse back to an unmagnetized state unless forced to do so. This phenomenon, which can be thought of as a kind of magnetic memory, is known as **hysteresis**. To correct for this nonlinearity, a high-frequency **bias** signal is added to the audio as it is recorded. The bias signal moves audio signals away from the zero magnetic point of the transfer curve to parts that are linear. The bias signal's frequency is ordinarily at least five times higher than the highest audio frequency to be recorded. Good-quality recording equipment typically uses bias frequencies in the range of 50 to 200 kHz. The bias signal is added as the audio is recorded to tape, but is discarded on playback because it only serves to keep the audio from being distorted by the recording process.

The physics of magnetism makes the ability to record audio evenly across the audio spectrum a problem. Generally, system output above about 5,000 Hz increases as the frequency rises. This is mainly caused by the physical properties of playback heads; their efficiency increases as frequencies rise by a factor of about 6 dB per octave. To standardize this effect so that recordings made on one machine can be played properly on another, an exact equalization curve must be tailored for each recorder. In the United States, the **NAB standard equalization curve** is used, but in Europe and most of the rest of the world, the IEC curve is employed.

Analog magnetic recording technology suffers from a variety of inherent noise problems, most notably **tape hiss**. Hiss is the noise that is created by the millions of microscopic magnetic particles on the audio tape passing the playback head, each one creating a tiny output for the playback circuit. A variety of solutions is available to combat the problem of tape hiss. Because most tape noise is concentrated in the upper portion of the audio spectrum, the simplest but least satisfactory of those solutions is to employ a low-pass shelving filter that reduces all the high-frequency content in the recording.

The **Dolby noise reduction system** has been at the forefront of noise reduction research since Dolby Laboratories' founding in 1965. This organization has produced a succession of noise reduction systems over the years, including Dolby A, B, C, S and SR, some of which are available for licensing by equipment manufacturers. The Dolby technique is described as **companding**, in which the overall signal level is compressed during recording and then expanded during playback. Concurrently, the high-frequency signal material is boosted during recording (or **preemphasis**), then cut during playback **(deemphasis)**. The final result is a sharp reduction in audible tape hiss without any serious degradation of frequency response. David Blackmer introduced the **dbx noise reduction** system in 1971, which operates on a similar companding principle. Although dbx technology never achieved great popularity for use on consumer equipment, it was widely utilized on small-format, multitrack tape machines aimed at the semiprofessional market.

ANALOG TAPE TECHNOLOGY

Analog tape technology still has a secure place in the recording industry. More than a few recording engineers and producers insist that analog tape offers a pleasant warmth that is missing from digital recordings. However, others maintain that this warmth is nothing more than **harmonic distortion** present in nearly any piece of analog equipment, but which is particularly evident in magnetic recordings. Whatever the case, it seems unlikely that analog tape will disappear entirely from the production scene anytime soon.

The major distinctions among the different conventional tape technologies such as reel-to-reel, cassette, and disk storage are in the kinds of recordings produced. Conventional tape technologies store information in **analog** form; a term derived from the fact that the information is analogous to, or like, the original sound wave. In other words, the patterns of the magnetic charges of particles on the tape are an analog of the electrical current supplied by the transducer. Tape machines differ in their **track** and **channel** configurations. *Tracks* are the magnetic paths onto which the signals are recorded on a tape, while the term *channels* refers to the number of inputs-outputs a machine has. The first tape machines manufactured had only a single channel that recorded just one track and were identified as **mono**, or monaural. This arrangement is also known as **full-track** mono, because the track occupies the full width of the magnetic tape. Later, after monaural tape machines, there came two-channel/two-track, or **half-track** stereo tape machines. These recorders laid two magnetic tracks, each of which occupied nearly half the tape width and were separated by a small blank strip called the guard band. The guard band reduced **crosstalk** between adjacent tracks. Other **multitrack** tape configurations eventually emerged, and included three-track, four-track, eight-, sixteen- and twenty-four-track; and again, all contained a like number of channels. As a rule, professional multitrack tape machines have the same number of tracks as channels, while some consumer tape machines and cassette tape machines are configured with four tracks and two channels. Figure 6.5 indicates the way tracks are placed on tapes in different configurations.

There is a direct relationship between track width and signal quality in analog recording, especially in regard to signal-to-noise ratio. As the number of tracks is increased, either the width of the tape must be greater or the width of the tracks will have to be reduced. Many of the earliest reel-to-reel machines used quarter-inch tape. The quarter-inch tape's width imposes limitations on the number of tracks that it can accommodate while maintaining professionally acceptable quality. Due to this relationship, as the number of tracks on professional multitrack

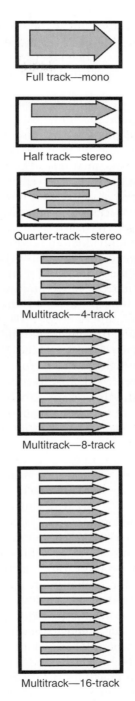

Full track—mono

Half track—stereo

Quarter-track—stereo

Multitrack—4-track

Multitrack—8-track

Multitrack—16-track

FIGURE 6.5   Magnetic recording-tape audio-track configurations.

recording machines grew during the 1950s and 1960s, tapes had to become wider. Four-track machines used half-inch tape, eight tracks used one-inch, and sixteen-track recorders used two-inch tape. These configurations provided the needed expanse of tape surface onto which signals could be recorded while retaining adequate guard bands.

By the 1960s, stereophonic recordings had become the norm, and **quarter-track** reel-to-reel tape machines were being produced for audiophiles as home stereo recording and playback systems. In order to maximize the program time on a single reel of tape and provide left and right channels, four interleaved tracks were placed on quarter-inch tape. The reel of tape could be played in one direction to its end, then the reels reversed by turning over the tape, and the second set of tracks could be played to its end. Unfortunately, the quarter-track configuration was made for small track widths and guard bands. Because of the reduced quality of recordings, the quarter-track stereo configuration was considered only a consumer format. The individual audio tracks on a quarter-track, quarter-inch tape are roughly 0.05 inches wide, versus 0.10 on a half-track, quarter-inch tape.

Advances in recording-head manufacturing processes in the 1970s and 1980s gave rise to narrow gauge multitrack tape machines. These were regarded as "pro-sumer" or semi-professional units aimed at the serious amateur market, but not up to the requirements of professional recording. Several manufacturers offered eight tracks on half-inch tape, and one company even developed a sixteen-track machine that used half-inch tape. The narrow guard bands produced recordings susceptible to high levels of cross-talk, and so these machines were never really suitable for professional work.

A second highly important variable affecting tape recording and playback quality is the speed at which the tape travels past the heads. **Tape speed** is specified in inches per second (ips) or, outside the United States, in centimeters per second. The faster the tape moves past the heads, the greater the signal-to-noise ratio and the wider the frequency response of the recorded signal. Professional tape machines move tape past their heads at 15 ips or, in some instances, even 30 ips. Consumer reel-to-reel tape machines marketed through the 1970s typically operated at $3\frac{1}{2}$ and $7\frac{1}{2}$ ips. The slower tape speeds permitted longer program times, but with a concomitant loss of fidelity. The slower tape speeds, coupled with the two-sided, quarter-track configuration, provided home recordists recording time on a 1,200-foot reel of tape sufficient to accommodate the program material found on a long-play album. Tape machines with $7\frac{1}{2}$ ips recording and playback speeds were used for spoken-word applications professionally, but rarely for music. Even slower recording speeds were available from a few manufacturers, $3\frac{1}{2}$ and $1\frac{7}{8}$ ips, for use as logging recorders purely to document audio events such as phone calls on 9-1-1 emergency lines or courtroom proceedings.

The Philips Corporation developed the compact cassette in 1963. The compact cassette uses eighth–inch tape for the quarter-track arrangement seen in Figure 6.6. Cassette machines normally operate at only $1\frac{7}{8}$ ips. Although originally intended for use in dictation machines, the cassette quickly gained popularity as a musical medium for the masses. Hundreds of millions of consumers know

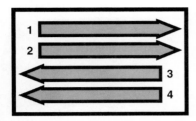

FIGURE 6.6   Compact cassette audio track arrangement.

the familiar compact cassette from their use in automobiles, home stereos, and portable cassette recorders through the 1980s and 1990s. Figure 6.7 shows one of the cassettes. Philips aggressively encouraged manufacturers to adopt their tape transport mechanism, and this contributed to its wide diffusion in the marketplace. Acceptance of the cassette as a music format was aided also by Philips' distribution of prerecorded Musicassettes through its Polygram division. Despite its popularity as a consumer format, the cassette has never been considered a professional tape format for music recordings. Its slow tape speeds and narrow track configuration are suitable for voice recordings such as on-the-spot interviews by broadcast journalists, but little else.

In recent years, the cassette mechanism has also found a niche in the small-format, personal multitrack market, with companies like Yamaha, Fostex, and Tascam employing the cassette tape and transport mechanism for their four-track machines. These small multitrack machines are intended exclusively for amateur music recordings and are marketed under names such as "personal multitrackers" or "portastudios." An example of one of these is shown in Figure 6.8. Miniature multitrack recorders like this one offer overdubbing and integrated mixing facilities. They are popular among composers to make quick recordings of song concepts, and despite their generally poor audio quality have occasionally been used to produce albums for commercial distribution.

FIGURE 6.7   A Philips compact cassette.

FIGURE 6.8   A portable multitrack machine that records on compact cassettes.

Contemporary analog tape recorders generally share common design archi-
tectures. Whether tapes are loaded on reels or housed in a cassette, the machines
used to record and play them back tend to have similar key features. These features can be placed into two categories: head arrangements and tape **transport**.
Heads are always arranged in the same sequence: erase, record, and reproduction
(playback). The layout of a typical tape deck is illustrated in Figure 6.9. This
arrangement allows a tape that contains a recording to be rerecorded. As the tape
passes under the **erase head**, the magnetic properties of the tape oxide coating
are scrambled, effectively erasing any existing material. New material can then be
recorded as the tape continues onward and passes beneath the record head. Some
machines also offer the opportunity to monitor the reproduction head as the tape
is recorded, thus ensuring that the material is being properly captured.

Head assemblies on multitrack recording machines are designed to allow
separate tracks to be individually switched between recording and reproduction
modes. In addition, multitrack machines incorporate facilities that allow the
record head for individual tracks to be used for playback. Machines like this permit a signal on one track to be played back via the record head, while the same
head is simultaneously recording new material on another track. Different manufacturers have distinctive names for this approach (e.g., "selective-sync," or
"selective-reproduction"). The purpose of this facility is to permit synchronization
of new material with existing tracks during **overdubbing**; any new material laid
onto the tape will be in precise alignment with previously recorded tracks.

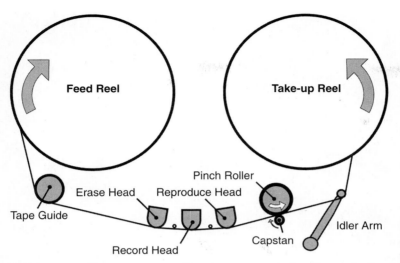

FIGURE 6.9   The arrangement of heads, capstan, pinch roller, and other parts of a reel-to-reel tape deck vary little from one machine to another.

Tape is moved through the machine by the transport mechanism. In most professional tape machines, three separate motors drive the **feed reel**, **take-up reel**, and **capstan**. The primary responsibility for moving the tape is assigned to the capstan and **pinch roller** assembly, while proper tape tension is maintained by varying the torque on the reel motors and by the arrangement of the idler arms. When the pinch roller and capstan are engaged, they are squeezed so tightly together that the tape moves at a rate totally controlled by the speed of the capstan's rotation. The photograph in Figure 6.10 shows the transport details of a typical reel-to-reel machine, and for comparison, Figure 6.11 presents a view of a representative cassette transport mechanism. The feed reel and take-up reel attempt to turn in opposition to one another, and thus keep the tape taut across the heads. Whenever the pinch roller and capstan are disengaged, and more power is applied to the feed-reel motor, the tape rewinds. If more current is applied to the take-up reel, then the tape moves fast forward. Reel-to-reel machines employ an idler arm attached to an automatic shut-off switch so that the drive mechanism halts when the tape runs out. Without tape tension to hold the idler arm up, it will drop and switch the motors off.

The **cartridge**, or broadcast "carts" as they are known, remain in extensive use among U.S. radio stations. Outwardly, the cartridge resembles the **eight-track cartridge** that enjoyed a brief phase of popularity as a home music format in the 1960s and early 1970s. Both types of cartridges trace their origins back to the four-track cartridge, developed by **Fidelipac** and championed by Charles Muntz.

Both the eight-track and the broadcast cartridges contain an endless loop of quarter-inch tape, which is lubricated by graphite or a similar powdered material. But that is where the similarities end. By endless, it is meant that the end of the

FIGURE 6.10   The transport and head mechanism of an Otari MTR-12 reel-to-reel deck presents a typical recording machine architecture.

tape was spliced to the beginning, so that the tape played continuously, starting at the beginning of the tape as soon as it finished. In the eight-track tape, extremely narrow guard bands separate four pairs of interleaved stereo tracks, thus the designation eight-track. The broadcast cart has only left and right stereo tracks, and a narrow track for a cue tone, providing enough track width to ensure reasonable

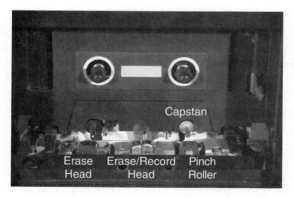

FIGURE 6.11   Compact cassette transport and head mechanisms are similar among most recorders.

sound quality. In the consumer-quality eight-track tape, a short piece of metallic tape is spliced into the tape at the point where the loop connects the tape end with the beginning. A sensor is triggered by the metallic tape to advance the head mechanism onto the next pair of stereo tracks. Instead of a metallic sensor tape, broadcast cart recorders place a tone signal on the cue track of the tape each time a recording is stopped and started. When the playback heads detect a cue tone—which is not audible in normal use—the machine stops the tape. By this means, the tape is always cued to the beginning of a recording cut and allows multiple cuts to be recorded on a single cartridge, with the succeeding cuts always cued for immediate playback. Carts are used primarily for the playback of commercial spots, but they have sometimes been used for music playback too.

Another significant difference between the eight-track and the broadcast cart is the location of the pinch roller. In the eight-track, the pinch roller was contained in the cartridge. This contributed to problems in seating the cartridge properly, resulting in inconsistent tape speed and consequent **wow** and **flutter**. The broadcast cart addressed the tape motion issue by putting the pinch roller in the machine, not the cartridge. Broadcast cartridges have an access hole in their bottoms to allow pinch rollers into the cartridge shell so they can engage the tape and the capstan. Differences between the two types of cartridges can be observed in Figure 6.12, which shows the broadcast cartridge on the right.

FIGURE 6.12   A cartridge used in broadcasting, on the right, has a large circular opening in the back where the pinch roller and capstan from the tape machine can enter. An 8-track cartridge on the left has a small opening for the machine's capstan, but has its own integral pinch roller.

DIGITAL RECORDING AND STORAGE

As outlined in the chapter on digital audio, storage of audio in the form of digital recordings entails the use of binary code, through which waveform values are represented by blocks of binary numbers made up of strings of the digits 0 and 1. Converting analog audio signals to digital data requires sampling by the analog-to-digital converter (ADC). The voltage arriving from an audio source is measured regularly and rapidly, and each measurement is registered in binary code.

Once converted to binary form, audio can be saved in storage media that use **laser** light beams to record and playback data. Light lends itself readily to binary coding because a beam is either on or off in the same way that digital information is either 1 or 0, that is, on or off. Recordings of this sort are termed *optical*; the most common are compact disks (**CD**) and digital versatile disks (**DVD**).

During the 1980s, audio CDs demonstrated the benefits of digital technology and became the first popularly adopted digital consumer format. In gaining such wide popularity, compact disks made analog phonograph recordings practically obsolete, though a small number of aficionados have kept the vinyl storage format alive. The CD format is small in size and reasonably durable and, most important, has superior sound quality compared to vinyl recordings. For example, CDs typically exhibit signal-to-noise ratios of more than 90 dB. CD players, even inexpensive ones, offer reasonably flat frequency response from 20 to 20,000 Hz, less than 0.01 percent total harmonic distortion, and wow and flutter too low to be measured accurately. Specifications of this level cannot be delivered by any analog recording system available today.

Each 4.8-inch compact disk can store about ninety minutes of uncompressed audio recorded as 16-bit samples at a sampling rate of 44.1 kHz. Digital data are recorded by forming microscopically sized pits in spiral tracks on the disk's recording surface. The pits represent the binary numbers that make up the samples. There are about five billion pits on a fully recorded disk, each pit representing a single digital bit. The spiral track is remarkably thin, and its total length can be as great as twenty-three miles. To appreciate the precision required in CD data-writing, consider that the length of a single data pit is roughly same length as a single E. coli bacterium. Each pit is approximately 0.5 microns wide and 0.83 microns to 3.56 microns long. A single human hair could completely obstruct tens of thousands of pits on a compact disk.

Compact disks store binary data in this manner: when the data state is "on," the laser turns on and burns a small pit into the surface of the disk. When the binary data is "off," the laser is off, and no pit is made. The area between the pits is termed the **land**, as is the area between record grooves in analog recordings. To retrieve data, a laser beam traces the spiral of pits as the disk rotates in a player, and the light reflects from the disk's mirrored inner surface to a photodiode pickup. Pits interrupt the laser beam as the disk rotates and the CD drive's electronic circuitry converts signals from the photodiode's output into the binary numbers of the samples. Data from CDs are retrieved from the bottom, so the pits appear as microscopic bumps. The bumps refract or bend the laser light of the CD playback mechanism so that the light is reflected back into a **photoreceptor** that converts light back into an electrical voltage pulse. When the light strikes the receptor, a current is formed; when no light is striking the receptor, no current flows. This process is illustrated in Figure 6.13.

To reduce the risk of potential errors in decoding pits impressed on a CD disk, a coding scheme called the **eight-to-fourteen modulation** method (EFM) is utilized. EFM shuffles the ordering of samples when recording to the disk so that if the disk becomes scratched, it is less likely that several samples in a sequence will be destroyed. If a series of samples is missing, the player will not be able to

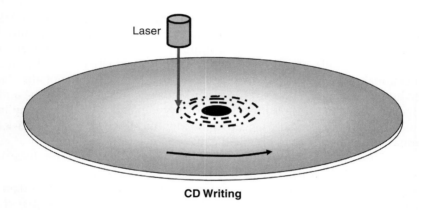

CD Writing

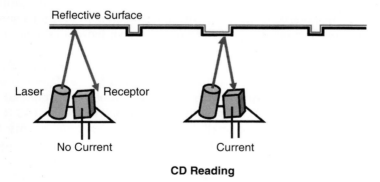

CD Reading

FIGURE 6.13   Top: To record a CD, a laser burns microscopic pits in the disk's surface. Bottom: To play back a CD, a beam from a laser is reflected into a photoreceptor thus generating an electrical output.

correct the errors. Of course, the proper order of samples must be restored on playing back the data.

In the mid-1990s, several manufacturers introduced professional standalone compact disk recorders, or direct-to-disk recorders. These machines record audio tracks directly to a compact disk in real time rather than by assembling the tracks on a computer and recording them on the blank CD in a single pass. Many stand-alone CD-Rs provide an operating mode that allows tracks to be added to the disk in separate sessions. A disk assembled in segments must then be finalized, or converted to Red Book standard, before it can be played in a conventional compact disk player. Once finalized, no more tracks can be added to the disk. Despite opposition by some recording-industry organizations, consumer-priced stand-alone CD-R models entered the market. To deal with the piracy concerns, equipment manufacturers are obliged to incorporate a copy protection scheme in these recorders, called the serial copy management system (**SCMS**). SCMS allows only one digital copy to be made from a compact disk. That copy cannot be digitally recopied, however, in a SCMS-equipped recorder. This prevents digital copying of commercially produced disks. Most consumer CD-R recorders and some professional models will only accept audio blank disks, and not the data disks used by computer drives for storing other kinds of data. The cost of the audio disks is slightly higher because it includes a royalty fee paid to the recording industry. Audio disks are also digitally tagged in the identification information track to mark them as audio disks and to disable digital copying.

In 1992 Sony introduced a technology that somewhat resembles the compact disk called the **MiniDisc (MD)**. The MiniDisc recording medium is a 64-mm disk enclosed in a plastic shell, and it was originally developed as a data storage format. The recordable MD uses a technology called **magneto-optical** recording, a technology related to the one used in rewritable compact disks. This method combines elements of magnetic recording with the laser techniques of optical storage systems. In magneto-optical recording, a laser selectively and precisely heats a small spot on the disk's surface, while the magnetic head simultaneously sets the magnetic polarity of the spot within the heated area. After recording, the disk surface cools, leaving the data permanently recorded until the laser reheats it again.

Several equipment manufacturers now manufacture MD players/recorders such as the one shown in Figure 6.14, but the MiniDisc has failed so far to gain much acceptance as a professional music-recording tool. The compression system used in MiniDisks is called Adaptive Transform Acoustic Coding (**ATRAC**). This compression scheme allows the small disk to contain the same seventy-four minutes of program material that a compact disk can accommodate. However, the compression ratio of approximately five to one yields a frequency response and dynamic range that is considered by many producers to be inadequate for professional and audiophile uses. Even so, MD recorders work well for location voice recording in broadcast newsgathering, and have maintained a market niche for consumer use.

FIGURE 6.14   A MiniDisc (left) and a MiniDisc recorder with a disk partially inserted.

DIGITAL TAPE TECHNOLOGY

The commercial launch of digital tape recording produced a myriad of types and configurations of tape-recording machines. Even though the technique of digitally coding sound by a computer was demonstrated in 1957, the first professional digital recording machines did not appear on the market for quite a few years. Sony and Studer Corporations collaborated to develop the **digital audio stationary head** or **DASH** format that appeared around 1980. As the name implies, the DASH machine used fixed heads over which the tape passed for recording and reproduction, and in that sense it looked like an analog tape machine. DASH machines were reel-to-reel, and were available in track configurations from two tracks on quarter-inch tape to forty-eight tracks on half-inch tape. In addition to Sony DASH machines, the Mitsubishi PRODIGI recorder was available for professional use through the 1990s.

There are, however, some noteworthy differences in the design architecture of the DASH machines as compared to an analog recorder. DASH machines require no erase head, so they have only two heads, and those heads are arranged in reverse sequence. That is, the tape passes over the reproduce head before the record head. The signal is encoded onto the tape in much the same way that digital data is stored on any other magnetic medium (as will be discussed in the following chapter concerning digital audio workstations). Stationary-head digital recorders have never truly gained the broad acceptance that other formats achieved, largely because of the high cost of the machines and the intensive maintenance regimen that they demand. Moreover, the rapid development of less expensive digital formats and the growth in digital audio workstations has contributed to the stationary-head recorder's small market share.

The problem with stationary-head technology is the difficulty of gaining sufficient bandwidth to properly record a fast-moving data stream. Several other stationary-head formats have tried to carve out a share of the audio market. The Sony Corporation offered its stationary [head] digital audio tape (S-DAT), a two-track digital cassette recorder, and the Philips Corporation introduced the **digital compact cassette** (**DCC**). These formats made use of a cassette similar to the compact cassette, but employed stationary record and reproduce heads. Although the DCC offered backward compatibility with analog compact cassettes, it never really achieved acceptance in the marketplace.

Most other digital tape-recording formats employ some form of cassette mechanism using **helical scan** techniques that use rotating recording and playback heads. Helical scanning gets its name from the spiral (or helix) wrap of the tape around the **head drum** that was common on some of the early videotape machines. Because the rotating heads spin rapidly, tracing tracks obliquely across the tape, the tape itself can move at a comparatively slow speed. The helical scan recording and playback process is illustrated in Figure 6.15. The helical scan system achieves a high head-to-tape velocity because of the rotational speed of the heads across the tape's surface as it moves past the head drum. The advantage of this approach is that helical scanning permits more data to be stored on shorter lengths of tape.

Perhaps the most successful of the helical-scan audio-recording formats has been the **digital audio tape** (**DAT**). The DAT format, jointly developed by Sony and Philips and unveiled commercially in about 1987, uses a unique small cassette, roughly half the size of a Philips compact cassette. The DAT cassette houses 4-mm (approximately eighth-inch) tape on which are helically recorded tracks containing digitized two-track audio. This system is more properly termed **R-DAT**, or rotary head digital audio tape, to differentiate it from the S-DAT stationary-head format just mentioned. The DAT digital storage scheme uses no compression, and while most professional recorders offer options of 32 kHz or 48 kHz sampling, most recorders use the same sampling and bit rates as compact disks—44.1 KHz, 16-bit. Because of the format's high fidelity and wide bandwidth, recording studios and broadcast stations quickly adopted DAT for a period, even though it was originally intended for consumer use.

The high quality of the DAT format fueled fears within the record industry over potential piracy problems. Therefore, the first generation of consumer DAT recorders did not allow digital copying; the only inputs they provided were analog. Subsequently, the serial copy management system (SCMS) described previously was integrated into nonprofessional DAT machines. While DAT achieved little adoption as a consumer format, it enjoyed acceptance among audio professionals for more than a decade. By the early twenty-first century, however, major manufacturers had dropped production of DAT machines, owing to the vast number of competing digital recording technologies.

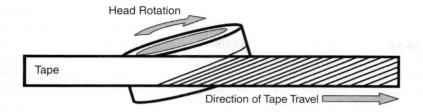

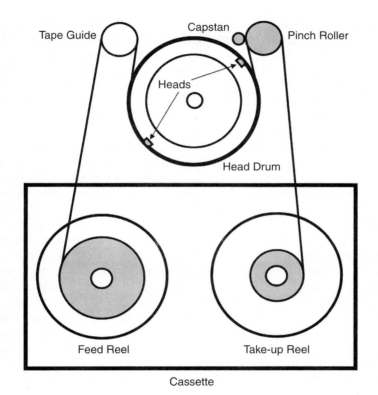

FIGURE 6.15   Top: The angle of the head drum in relation to the tape creates slanted electromagnetic tracks on a helical tape recording. Bottom: The layout of a helical recording mechanism for cassettes requires the tape to be pulled from the cassette and wrapped around the head drum.

DAT recordings can be stereo or, rarely, monophonic, but a class of multitrack digital tape-recording formats often referred to as a digital tape-recording system (**DTRS**) was popularized in the 1990s. Recorders using these formats incorporated a rotating head and cassette configuration similar to the DAT. For example, Tascam's popular DA-88 series of digital multitrack recorders used the same 8 mm cassette tapes intended for 8 mm video camcorders. Alesis and Fostex also manufactured digital multitrack recorders, using **ADAT** (Alesis Digital Audio Tape) format that utilized standard Super-VHS videotape cassettes as the recording medium. Both types of multitrack digital machines operated in much the same manner, and from the operator's perspective offered the same functionality as their analog counterparts and, most important, synchronized overdubbing. A great advantage offered by these low-cost DTRS multitrack machines was their modularity. Multiple machines could be synchronized together to expand track capabilities so that up to twelve eight-track machines could be stacked together to permit ninety-six-track recording.

Professional digital hard-disk multitrack recorders, or **HD** recorders, came into general use in the 1990s, and then slightly later they arrived in the semiprofessional market. As the name suggests, hard disk recorders store digital audio on the same kind of hard disks found in computers. Disk-based recording offers the advantages of random accessibility to recorded material and security of long-term storage. As in the computer hard drive, any data may be instantly retrieved from the disk. This is quite unlike tape-based digital recorders that need fast-forwarding or rewinding to locate needed recordings.

HD recorders are capable of nondestructive editing, through which editing is accomplished by rearranging the sequence of data retrieval. For instance, if the operator wishes to cut a thirty-second passage from a recording, the recorder's software is simply instructed not to play back the passage, rather than actually cutting it from the recording. Because the source recordings are not altered in any way, edits can be undone at any time.

HD recorders are available as stand-alones, as computer peripherals, or as part of packaged mixer/recorders, sometimes called desktop systems. As a stand-alone, the HD recorder is most commonly configured to provide a twenty-four-track capacity. Similar to DTRS and ADAT recorders just described, HD recorders are stackable to allow a practically unlimited number of tracks. Some HD recorders offer one or more hot swappable hard drives. Hot swapping or hot plugging means that as drive space becomes filled, an individual hard-drive assembly can be easily removed and replaced with another while the machine is still operating. No rebooting of the system is required.

SOLID-STATE RECORDING AND STORAGE

The most recent entry in the recording equipment field is the solid-state recorder. With no moving parts but having user-defined compression ratios and extended recording times, solid-state recorders are proving to be most popular for field recording but have the potential to be useful in studio recording. A number of manufacturers (Marantz, Panasonic, Sony, Tascam, and M-Audio, among others) offer recorders that make use of a variety of solid-state memory formats. Based on storage technologies commonly known as flash memory, their popularity has grown quickly and their cost has declined steeply. Solid-state storage has evolved into a wide variety of formats known under trade names such as Compact Flash, Smart Media, Memory Sticks, and Secure Digital (SD). Solid-state storage is a form of **EEPROM** (electrically erasable programmable read-only memory) in which data is stored as images of electrical charges.

A solid-state storage device is constructed of arrays of transistors, each of which has two control terminals—one called the control gate and the other the floating gate—that are separated by a thin oxide layer. The floating gate is totally surrounded by the insulating oxide layer, thus electrons that are forced onto it cannot escape. The presence of these electrons (or not) in these transistor cells is the basis of data storage in flash memory. Data are read by supplying a voltage to the control gate and checking the current flow. The flow depends on whether electrons have been stored in the floating gate. The construction of memory cells can be seen in Figure 6.16.

As shown in this drawing, when the current sensor detects a voltage on the floating gate, the result is interpreted as the number 0, but if the flow drops sufficiently, the sensor interprets the stored value as the number 1. Internal circuits within the flash memory module route electrical current to different sectors to read, write, or erase memory blocks. A new blank or unrecorded memory device contains cells all having the value 1. Erasing a solid-state memory device requires reopening all the gates. The erasing voltage may be routed to individual sectors of the storage device, to erase specific blocks, or to the entire device.

Perhaps the greatest advantage of flash memory is that solid-state storage involves no moving parts. Because of this, they are inherently more reliable than

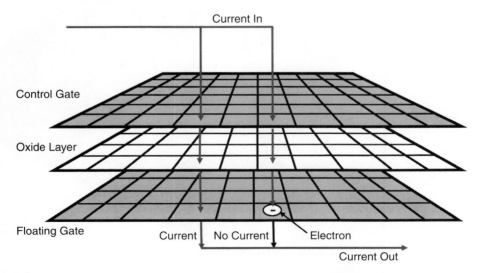

FIGURE 6.16   Solid-state storage employs layers of microscopic cells whose electrical charge (on or off) is determined by a single electron. The pattern of charges can store binary data.

types of storage that rely on mechanical processes. A second advantage of flash memory is that they do not need power to retain data. And once recorded, these devices keep data indefinitely.

Flash memory is the method employed for storage of audio recordings in many MP3 players, such as the **iPod** Nano and Shuffle. Solid-state storage devices are also used in recorders intended for professional portable use. Professional flash memory equipment has replaced DAT recorders for field recording in many situations.

The majority of solid-state recorders permit the user to select from a range of compression ratios that vary from no compression to aggressive compression ratios typical of MP3. It goes without saying that as the compression is increased, the recording time available increases correspondingly. The result is that a one-gigabyte memory card may hold as much as thirty-six hours of heavily compressed (MP3 at 64 kilobytes per second, or KBps) monaural audio. Only about thirty-five minutes of uncompressed WAV files recorded at 16-bit 44.1 kHz will fit on the same card. A great advantage of solid-state formats is that audio files can be transferred directly to a workstation via a card reader, rather than being played back and rendered in real time. This dramatically lessens the time required for transferring lengthy recordings. Also, files on memory cards can also be easily copied to other storage devices, transferred across the Internet through file transfer protocol (FTP), or attached to email messages. As the storage capacity of solid-state devices has increased and their physical sizes decreased, recorders using them have become smaller and lighter too. Figure 6.17 shows a professional solid-state recorder. Pro-

FIGURE 6.17 A solid-state recorder and media card.

FIGURE 6.18   A microphone with integrated solid-state recorder.

fessional recorders like the Flash Mic from HHB integrate the solid-state recorder into a handheld microphone, as shown in Figure 6.18. Consumer devices like the Apple iPod have been shrunk to the size of a package of chewing gum.

Data-storage technology continues to evolve as this is being written. For instance, magnetic random access memory (**MRAM**) is poised to supplant existing solid-state storage methods. MRAM makes use of electron spin to store information. This technology is somewhat like flash memory in that it does not need power to retain recordings, but MRAM requires much less power to read or write data. Magnetic RAM thus could extend battery life in portable equipment. Magnetic RAM shares architecture similar to other types of solid-state storage, but its memory cells change in magnetic polarity to represent the ones and zeros, rather than store an electron charge. MRAM also has the potential to read and write the data faster, and to store more data than in conventional flash-storage methods.

CHAPTER SEVEN

DIGITAL AUDIO WORKSTATIONS AND COMPUTERS

Because of the audio field's transition to digital technology, a large portion of recording and editing now takes place on digital audio workstations rather than on reel-to-reel or cassette-based recording equipment. A **digital audio workstation (DAW)** is a sound recording and playback system built around a computer. There are two general types of DAWs, one based on standard personal computers and the other based on proprietary designs, often called desktop **HD recorders**. These two types function similarly but make use of different kinds of hardware.

Because of the importance of digital audio workstations in contemporary audio production, this chapter will present information on their design and operation. Knowledge of how computers are constructed and how they function is vital for audio engineers and producers. They should be able to use computers with ease and be able to diagnose and correct problems in their operation. Although a complete treatment of the subject of computers is beyond the scope of this text, the following pages present an introduction to basic elements of digital technology that are relevant to the audio field. First, a few notes on the different sorts of DAWs that are available.

DIGITAL AUDIO WORKSTATIONS

Desktop recorders are the digital descendants of the cassette tape–based multitrack recorders popular in the 1980s, and they offer the capability of a simple studio within a single "box." Desktop HD recorders usually integrate all the functions required to convert an audio signal to digital data, to process, store, edit, mix, play back, and output the signal. Most desktop recorders are stand-alone units that physically resemble small audio mixers. These recorders have microphone preamplifiers, A-D and D-A converters, software-based effects processors, an internal hard drive, simple display screens, mix-down facilities, and a CD burner for output. These units have familiar interfaces, including tactile features such as faders and knobs, so many recordists find that they require little in the way of specialized skills to achieve good results. Desktop recorders may provide eight, twelve, sixteen, or even more tracks. Microphones or line-level inputs can be connected directly to the desktop recorder; their signals can be digitized, stored, equalized, edited with effects added, and then mixed down for recording on the internal compact disk drive.

Desktop systems employ proprietary hardware designs and can be used only as DAWs—they are not general-purpose computers. Often also termed "all-in-one" DAWs, there are many examples of proprietary systems, such as ones manufactured by Roland, Yamaha, and Tascam. Systems like these are especially popular in semiprofessional and project studios. The discussion in this chapter mainly concerns DAWs constructed around conventional IBM-PCs and Apple Macintosh computers that have become de facto standards in professional studios. However, many of the points made about this kind of workstation can be applied to the desktop DAW as well.

WORKSTATION ARCHITECTURE

As just noted, when equipped with suitable audio processing hardware and audio recording and editing software, a computer becomes a digital audio workstation or DAW. Systems like this record and play back audio files **"direct to disk,"** that is, on the computer's own hard disk where they can be easily processed and stored. Since most audio is likely to be digitized anyway, transferring data to a computer for storage on its hard disk will be simple and quick. And once stored in the computer, there are many software applications that can edit, clean up, or otherwise process the digital audio files. Software is available not only for professional audio applications but also for semiprofessional and consumer uses as well. Often, even modestly priced software can operate satisfactorily on computers considered slow by present standards.

In professional use, a system built around a conventional computer employs specialized hardware containing ADC and DAC circuits. This hardware may include components external to the computer itself. Although most home computers contain sound cards that achieve similar results, professional equipment is designed to meet higher technical standards and to provide more flexibility. For instance, professional equipment commonly provides for multiple independent input channels, whereas home computers provide stereo input channels only. These professional units also have other desirable features including multiple outputs, enhanced signal-to-noise ratios, and lower distortion specifications.

As an example, the Digidesign Corporation has produced a range of hardware products to support its Pro Tools editing software. These include specialized cards that can be inserted into computers and that connect with external interface units. Audio is sent to the interface for processing and conversion into digital data, and from the interface the digital stream is forwarded to the computer. Other manufacturers provide comparable hardware and software options.

Another digital audio workstation layout employs hardware devices called control surfaces that are attached to the computer and closely tied into its system by the use appropriate software. Like the semiprofessional desktop recorder, workstations incorporating these devices mimic the operation of analog mixers. Professional versions of integrated systems like this were particularly popular at a time when most producers were more comfortable with equipment that resembled analog systems. Setups like this had the added benefit of reducing demands on the computer's processing capability by use of the control surface's own microprocessor. Control surfaces eliminate or minimize the need to use the keyboard and mouse to operate the system computer. However, these systems add expense to construction of a production facility and the greater processing power of modern computers has reduced the need for processing help, so today the use of control surfaces is usually optional.

The precise arrangement of hardware components in a computer is commonly termed its *architecture*. This is important because the construction of computers greatly affects the way they can function as audio workstations. In the first few years following personal computers' appearance in retail outlets, there were many different types available. As time passed, market forces gradually reduced the number of computer types used in most settings to only two—the IBM-PC and the Apple, later the Apple Macintosh. Personal computers of the PC type, that is, ones that can use the Microsoft Windows operating system, are dominant and they comprise more than 90 percent of computers in use today. Even though the Macintosh has been favored in media and the arts, and even though Apple has enjoyed commercial success with its Macintosh line of computers, its market share is small. There is a third personal computer that has a small number of users—the open-source Linux operating system. It has few adherents among media professionals but is popular in specialized settings such as Web servers.

Figure 7.1 shows how different subunits of a personal computer are connected in a typical IBM-PC computer. The architecture of the PC presented in this block diagram is simplified to highlight its key components. The "brain" of a personal computer is its microprocessor, which is responsible for the calculations that must be performed to process data. It carries out actions, step-by-step, following instructions provided by software programs operating in the computer. Mainly, these instructions direct the microprocessor to carry out three functions: to perform mathematic calculations, to make decisions based on prescribed instructions, and to control movements of data throughout the computer. The microprocessor is contained in a single module that contains the entire central processing unit, or CPU. The microprocessor is an exceptionally complex piece of electronics. For instance, the Pentium 4 microprocessor contains about 125 million transistors. The CPU carries out instructions contained in software, each instruction defining a single small step of data processing.

The number of different instructions a microprocessor can recognize is usually quite limited, not more than one or two hundred, depending on its design. The ability of a microprocessor to carry out instructions rapidly is critical. Even in ordinary home computers CPUs can complete millions of instructions every sec-

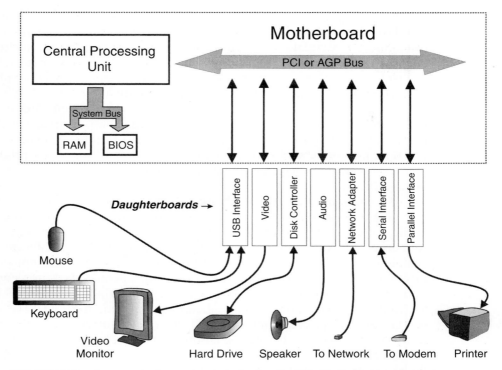

FIGURE 7.1   A computer is constructed of subunits. This block diagram shows how a typical IBM-PC might be built from a motherboard and a number of daughterboards.

ond. The faster the microprocessor, the more data it can process in a given period of time. Audio workstations offering a capacity of twenty-four or more audio tracks demand powerful microprocessors in order to keep up with the massive data flow. One of the most important sets of instructions allows the CPU to move data in and out of storage locations. Computers use two kinds of memory to store data: temporary, short-term data storage in memory that can retain data only while the computer is operating (**RAM**) and long-term storage such as hard drives.

The microprocessor also contains a section devoted to control functions, primarily bus management. A computer bus is an electronic pathway through which data is routed to different sections of the system. The microprocessor monitors the movement of data on the bus and manages distribution of data and control signals throughout the computer via buses. In PCs, the arrangement of buses varies, and numerous types have been used under different nomenclature. At one time the ISA bus was universally used, but later the PCI became popular. **AGP** and SCSI buses have enjoyed popularity too. Differences among buses lie in the data formats and speeds and in the protocols employed for data movement. Although on the whole quite similar, bus types are sufficiently different that equipment designed for one cannot be used on another.

In desktop versions, both the PC and the Macintosh make use of a motherboard construction. A motherboard is the main printed circuit board of a computer containing its essential components, such as the CPU and associated circuitry. In order to make the computer fully functional, additional hardware must be added via banks of connectors. For instance, video outputs can be added by plugging a printed circuit card into a connector—or slot—on the motherboard. When a card is attached by plugging it into one of the slots on the motherboard, it connects with a bus of the computer, typically a PCI bus. The main advantage of a motherboard and slot design is that the configuration of the computer can be modified at any time merely by adding or removing cards. The selection and installation of these components largely determine the functionality of a computer intended for use as a DAW.

An enormous variety of cards are available to customize equipment for specific uses. For example, many audio cards are available for professional use including ones offering S/PDIF, AES, FireWire, SCSI, or USB interfaces for connection to external devices such as storage media. Other audio cards accept analog input and perform conversion to digital formats. To accommodate these cards it may be necessary to install drivers for the operating system, as will be explained shortly. Normally, computers that fit the description of a DAW are "desktop" systems, but with the right peripherals, laptop computers can serve as DAWs, too. Of course, laptop computers and all-in-one computers such as the iMac normally do not have the ability to be customized through add-in cards. These computers have to be adapted by the use of external equipment via USB ports or, in the case of laptops, PC Cards.

DATA STORAGE

As mentioned earlier, RAM is used for short-term, temporary storage in a workstation. RAM is an acronym for *random access memory,* so called because this sort of memory is arranged according to "addresses" and can be accessed in any desired order—in other words, randomly. This is in contrast to serial types of digital storage, such as tape, in which data can be accessed only in sequence. The addressing of RAM works like house-numbering systems, with each storage location assigned a unique numeric locator. Of course, RAM is useful only as long as power is applied to the computer. As soon as a computer is switched off, all data stored in RAM is lost. To overcome this problem and to provide long-term storage, computers use magnetic hard drives where no power is needed to safely retain data. The volatile nature of RAM is familiar to anyone working on a computer file when power is unexpectedly interrupted. If this should happen, all work since the last hard drive save will vanish along with the data being stored in RAM.

Because RAM storage is volatile, it is important to have a reliable means to offload files for long-term storage. Computers usually accomplish this via magnetic recordings that are not dependent on power sources, most commonly an internal or external hard drive. To retrieve files, data are loaded from hard drives and then sent through a bus to the primary memory and to the microprocessor. Computer files containing audio data can require enormous hard drive space. Although audio files are most often stored on a computer's internal hard drive, they may also be stored on an external drive. In addition to the advantages of the non-linear, nonvolatile storage already described, external drives offer the advantages of high data capacity, interchangeability and portability, and—if high-speed interfaces are employed—fast read-and-write speeds.

The term *hard drive* came into use as a way of distinguishing this form of storage from the *floppy drives* that make use of removable disks. The magnetic recording medium inside a floppy disk is manufactured in much the same manner as audio tapes described in the preceding chapter. However, the shape of the recording medium is different—a thin circular disk instead of a long ribbon. Both floppy and hard drives store data in magnetic form so that data recordings may be readily erased and rewritten. Like RAM, these drives locate data according to unique storage addresses, so that data can be retrieved in any desired order. Floppy disks have become obsolete as storage media; their capacity is too limited for present demands, they are slow, and not always reliable.

A hard disk or hard drive, sometimes also known as a fixed disk, like other magnetic recording technologies, is constructed from oxide bound to a substrate surface, but in this case the substrate is made of a circular polished glass or aluminum plate. A hard drive contains one or more separate platters attached to a common spindle, with each platter providing magnetic recording surfaces on its top and bottom, as shown in Figure 7.2. Hard drives employ tiny recording and playback heads fixed on levers that move across the platters to locate data recorded in storage addresses. The head is moved to desired tracks on the platters by a mechanism that permits the heads to glide close to the magnetic surface without touching it. Figure 7.3 provides a closer look at the head assembly used in hard drives.

Before use, all disks must be formatted. Formatting writes reference points to the drive so that the computer can find addresses on the disks. Hard drives are somewhat fragile because the heads must be placed close to platters when reading

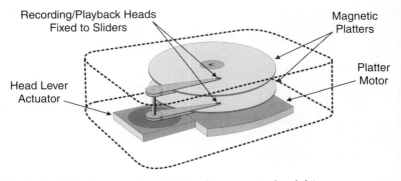

FIGURE 7.2   Construction details of a computer hard drive.

FIGURE 7.3   Computer hard drive head assemblies.

or recording data but they must never touch disk surfaces. The construction of the drive provides protection against physical shock that might cause a platter-to-head collision, but a crash is still possible if a drive is subjected to movement while operating. Because the hard drive presents a risk of failure, it is always prudent to make frequent backup copies of work as a precaution against crashes.

CD drives represent another kind of permanent storage used in workstations. Compact disks can be used to store audio either in the form of data or as an audio disk. When saved as data, audio may be recorded in a variety of formats—as a Windows PCM (.wav) file, as an audio interchange file format (.aif), or as a Windows media audio (.wma) file, among many others. When CDs are employed to store data, the files appear to users as they do on any other drive, but beneath the surface, the way that data are stored is specific to the CD storage mode.

An additional optical disk-storage medium is the DVD, the digital versatile disk. As noted in the preceding chapter, it is referred to as "optical" because DVD drives—like CD drives—use laser light beams to record and retrieve data from disks. In practice, the DVD is simply a medium that holds more data and can access data faster than the CD. Whereas the CD is capable of about 800 megabytes (MB) of data storage, an ordinary DVD recording can hold 4.7 gigabytes (GB) and the newer, double-layer disks can hold 8.5 GB. The DVD therefore possesses up to ten times the storage capacity of a CD. Because their capacity is large enough to permit storage of high-quality video recordings, they have become the preferred medium for movies, replacing the videocassette for most purposes. They have also made inroads on CDs as storage media, and will likely replace them altogether eventually. Adoption of DVDs grew rapidly following their introduction in 1997, and by 2003 there were 250 million playback units in use around the world.

The DVD offers an excellent storage medium for audio data because of its huge capacity and low cost; workstations commonly incorporate a DVD drive for archiving purposes. The audio DVD, known as DVD-A, provides seventy-four minutes of ultra-high-quality audio at sampling rates up to 192 kHz. The audio may offer full surround sound—that is, six channels instead of just two in stereo—and disks may also provide interactivity, contain video clips, and present other special features. Typically, DVD-A recordings use a 92 kHz sampling rate with 24-bit samples. Although there is enthusiasm among professionals for this new audio format, it is not clear how wide the consumer acceptance will be for it. In addition, a competing format known as the Super Audio CD has been introduced by Sony/Philips. This format provides multichannel capabilities, usually in the form of surround sound, as well as higher-quality audio. The Super Audio CD also incorporates content protection schemes that make duplication on ordinary CD drives difficult. Drives manufactured for this format have the additional advantage of backward compatibility with conventional audio CDs.

The format used to store data optically depends on the kind of disk employed. The DVD-RW is a disk that can be written and rewritten somewhat like a hard drive, making it particularly useful for archiving purposes. DVD-RAM disks are also like hard drives, but not so commonly used. For compact disks, standards are laid out in documents called "books" for each type of CD. For conventional audio CDs, specifications are spelled out in the Red Book, and for CD-ROMs containing data they are laid out in the Yellow Book. These detailed technical guides ensure compatibility among all players and computer drives and have been adopted worldwide by manufacturers.

INTERFACING DAWs

The term *interface* is used to describe devices and protocols that link a DAW with the world outside. Some interfaces receive data from external sources, other interfaces send data outward to external devices, and some do both. A workstation utilizes interfaces to communicate with users by means of a mouse, a keyboard, and a video monitor or possibly a control surface. To connect these devices with the computer, different types of interfaces have been employed, such as the PS-2 for the keyboard and mouse and the 15-pin VGA connector for computer monitors in PC-type computers. Older Macintosh computers utilized a completely different set of interface standards such as the ADB for keyboard and mouse and a "high-density monitor port" for video output.

In addition, full-sized PCs may include all-purpose interfaces such as the serial and parallel interfaces. These are sometimes termed *legacy* interfaces or *ports* because they are ones that have been traditionally included in workstations, though today they are often superseded by more powerful and flexible types. Both the conventional serial port and the parallel port have largely been supplanted by the USB. The USB—or universal serial bus—has become a standard used for connecting practically any kind of peripheral with a computer. Printers, keyboards, mice, modems, and all sorts of storage media now often connect to PC and Macintosh platforms via USB ports. Computer operating systems can, in principle, manage up to 127 different devices attached to it by USB, though as a practical matter most computers offer only a few connector ports. USB connectors are compact and are easily attached to computers. If a user wishes to connect more devices to the computer than there are jacks, it is possible to employ a hub—a device that has a single input and multiple output jacks.

Compared with old-fashioned serial and parallel ports, the USB enjoys a number of distinct advantages. First, the USB is designed to supply a small amount of power to attached devices. Some small external hard drives can be operated from USB ports, though peripherals that require greater power such as printers must have their own power supply. A second advantage of the USB interface is its throughput capability. The second-generation USB, known as USB 2.0, can manage a flow of data many times faster than a conventional serial port. At this speed, it may even be possible to use USB ports for interconnection of hard drives in audio production systems. Finally, most USB equipment is hot-swappable, meaning devices can be attached or disconnected from computers without rebooting. Because of these important features, the USB has replaced most other types of interfaces, except perhaps the FireWire interface.

The IEEE-1394 interface, often called FireWire (the name adopted by the Apple Computer, which popularized this standard), is a favored standard for use in professional studies and it is the predominant choice for connecting video devices with computers. Like the USB, IEEE-1394 interfaces can be used for connection of a wide variety of devices at a very high rate of data flow. Also, like the USB interface, IEEE 1394 ports support hot-swapping and may provide a limited amount of power to peripherals attached to the computer. There are two types of connectors used for FireWire ports, 4-pin and 6-pin jacks, with the latter connector capable of supplying power to connected devices.

Before either the USB or IEEE 1394 interfaces became popular standards, the chief means of high-speed connection between a peripheral and a workstation was by means of the SCSI standard. **SCSI**, or the small computer system interface, is still in used in professional audio workstations, but not so commonly as before. Early Macintosh employed SCSI to connect internal drives as well as external devices. Over the years, the SCSI standard has become diversified with "wide" and "narrow" variants of various throughput capacities. SCSI devices must be given unique identification numbers and multiple devices can be connected together in a chain but one of these devices must be "terminated," that is, shunted by a **resistor**.

One final interface that had been often used in audio production for transfer of audio files is the **S/PDIF** (Sony/Philips digital interface), though it was used only to transport digital audio between source and destination. It cannot be used to connect peripherals to a workstation. Digital outputs from DAT machines usually fit this standard, and other kinds of equipment (mixers and other sorts of recording devices) sometimes used it as well. Two connectors are associated with S/PDIF, the RCA coaxial connector, as commonly used in consumer analog audio equipment, and a unique optical connector attached to fiber optic cable called TOSLINK. A similar digital audio standard is the AES/EBU (Audio Engineering Society/European Broadcasting Union) interface. The AES/EBU interface normally employs 3-pin XLR connectors, just like ones used for professional microphones.

DAW SOFTWARE

So far, this discussion has concerned only hardware, that is, the mechanical and electronic components that make up a workstation. However, only when coupled with software—the nonphysical portion of the computer—will the workstation be able to carry out useful functions. Most audio software applications share similar screen interfaces. A typical screen view is shown in Figure 7.4, offering a view of the audio waveform for editing purposes, tool bars for editing, signal processing and effects, and a set of transport controls for playing, fast-forwarding and rewinding. This view is of an Adobe Audition screen; additional software production suites are illustrated in the following chapter.

As described earlier, software is made up of sets of instructions that tell the computer's hardware what to do, with these instructions arranged step-by-step.

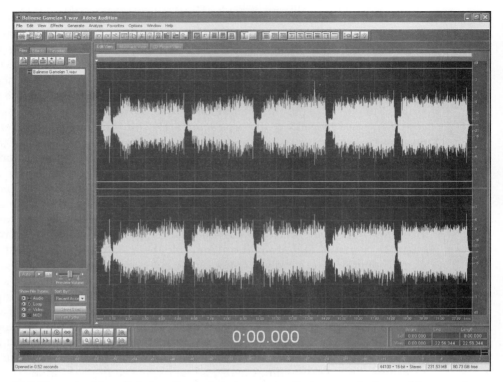

FIGURE 7.4   Most digital audio workstation screen displays are similar to this example. Here a stereo recording provides two waveform displays at the center, with the timeline just below. Player controls are at the lower left, and at the top are drop-down menus. Just below the menus are customizable shortcut buttons.

Software is normally conceived of as fitting into two distinct categories: operating systems (Windows, Macintosh OS, or Linux) and applications (such as audio editors, spreadsheets, or word processors). The following chapter will deal in detail with the application software used in computer-based DAWs, but a general examination of other kinds of software and firmware needed in workstations is appropriate here.

Before either operating systems or applications can function, there are two other kinds of instruction sets that a workstation requires: BIOS and drivers. The term *BIOS* is an acronym that stands for basic input-output system, a list of steps used to boot up the computer and to organize the flow of data from the CPU to peripherals and computer components such as the keyboard and mouse. The BIOS instructions are contained in a module on the motherboard where they are permanently recorded so that they can be available immediately at computer startup. BIOS are intended to present uniform conditions for the proper functioning of an operating system. BIOS modules must be individually tailored for the hardware items contained in the system. Because the BIOS employs instruction code it is like software, but because it is contained in a permanent integrated circuit chip it is like hardware; thus, the term commonly applied to BIOS is *firmware*. The operating system, relying on information passed to it by the BIOS, takes over operation of the computer. The operating system need not be aware of specific details of the computer's hardware because the BIOS handles communication with items in the system.

To help computers understand how to interact with attached equipment, operating systems and BIOS depend on drivers. Drivers are small bits of software that contain translation information necessary for instructions to be properly interpreted by hardware on the system. When Windows or Macintosh OS wants to access a piece of hardware incorporated in the hardware system, the operating system moves the drivers into memory and calls on the software routines contained in the driver instructions to carry out desired actions. When operating systems are installed, a database of common drivers is installed along with the software. When a new piece of equipment is added to the computer—for example, by connecting to a USB port—the OS examines the item to determine what it is and to ascertain what kind of driver it needs. If the OS fails to find a suitable driver already installed, the user will be prompted to provide it.

The foremost aim of the operating system is to present a consistent software platform so that application programs need not account for differences from computer to computer. Any application software should function precisely the same no matter what sort of computer it is installed on. For example, a computer might have a mouse connected via a PS-2 or an USB interface—both types work equally well with any late version of Windows—and software needs to operate correctly with either. Audio editing software is completely unaware of which type of mouse is installed, because it uses the operating system to control hardware attached to the workstation. Although this is a simple requirement, the goal is not easy to achieve in functional operating systems because the range of hardware possibilities is essentially unlimited. An operating system's quality is judged by how smoothly and transparently it handles this task.

Operating systems used in most audio production settings are single-user multitasking types. That is, the operating system expects input from one user at a time (although at different times there may be different users) who may have multiple software applications running simultaneously. Provided that the processing capacity of the computer is sufficient, multitasking operating systems allow the user to, say, record audio to a hard drive while at the same time browsing the Internet. Both Macintosh OS and Windows fulfill requirements of this type of operating system.

To understand how these operating systems evolved, a little historical background may be useful. The IBM-PC was first offered for sale in 1981, built to use an existing operating system called MS-DOS (Microsoft disk operating system). IBM renamed this operating system PC-DOS (personal computer disk operating system), and thanks to the quality of the hardware and IBM's marketing power, this operating system became an industry standard. Like other operating systems of that period, PC-DOS was text oriented; users sent instructions to the operating system through the keyboard while letters, symbols, and words were displayed on the computer monitor. As such, it required the user to communicate with the operating system using arcane commands and to interpret equally opaque signals sent to the screen. It soon became evident that many people had an aversion to computers because of the steep learning curve needed to become expert in the mystifying world of DOS.

Apple Computer, which had its own text-based operating system, also developed graphically oriented operating systems. A graphically oriented operating system uses pictures or "icons" to represent application software, and rather than using text commands it relies on commonsense language to communicate with users. In 1983 Apple released the Macintosh computer, which used a graphically oriented operating system. This Macintosh OS won acceptance among professionals working in the graphics and media arts, and the Mac's popularity grew as successive releases of the OS added color capability and as powerful professional-quality software became available. The single event that raised Macintosh's standing among audio producers was the introduction of the Pro Tools audio editing system. Initially, this package was aimed only at professionals working in production settings and was available only for the Macintosh computer. Pro Tools was not only the first truly professional recording and editing system but through the years has maintained its position as a preeminent professional audio production suite.

Microsoft soon recognized the benefits of a graphically oriented operating system and released the first version of Windows in 1985. This operating system continued to evolve through additional releases such as Windows 95, Windows 98, Windows NT, Windows ME, Windows 2000, and Windows XP. An array of capable professional audio production packages is now available for the Windows operating system, including a version of Pro Tools.

Finally, on the subject of operating systems, some mention must be made of UNIX and Linux, even though very few professional workstations in audio production use them. These two operating systems—and some other related OS—are quite similar, enough so, in fact, that the same applications generally work on both. Their main importance in the media field is in networking. These are particularly robust operating systems and both provide excellent security provisions, something essential for any computer used in a networked professional environment and especially when the computer is accessible via the Internet. Linux differs from UNIX in that it is an "open source" operating system, meaning that its code is publicly published and is constantly being improved by thousands of volunteer programmers. Linux is freely available and versions for both Macintosh and PC computers can be downloaded without charge. It is worth noting that the latest version of the Macintosh operating system, OS X, is built on a variant of the UNIX system and, thus, benefits from the enhanced stability and security provisions of that operating system.

CHAPTER EIGHT

AUDIO EDITORS
AND EDITING

Audio editing is the process of selecting and organizing recorded audio elements to achieve an effective and pleasing sound. Editing allows the producer to choose the best portions of recordings and to arrange them into a finished package, eliminating audio components that would diminish its overall impression. Editing can also be used to tighten a production by quickening its pace or to trim the running length of a production.

Digital audio workstations (DAWs) make editing easy and accurate by providing the cut-and-paste convenience of word processors and by enabling precision in edit point selection. Any errors that are made can be corrected easily, and edits can be undone easily. For these reasons, DAWs have made audio editing a much more pleasurable task than it was in the days of tape editing when editors marked edit points on audio tape with a grease pencil, cut it with a razor blade, and then finally joined the remaining sections of audio tape with splicing tape.

Due to the elegance and enhanced power of contemporary audio workstations, editing has become more widely practiced among professionals, and this in turn has made editing skills more important for all kinds of productions. In working with previously recorded music, an **editor** may insert special percussion effects—for instance, reverse cymbals—moving them to positions within the final mix that achieve a more pleasing result. Those who work in DAW-equipped radio or advertising production studios now no longer have to record their copy in single, uninterrupted takes. Instead, personnel have the luxury of selecting the best sections of multiple takes, which they may readily compile into one master commercial, news package, or other type of production. In this way, DAWs can be used to add polish and artistry to nearly any kind of audio production.

Digital audio workstations afford editors enormous creative possibilities. Unlike tape recordings, multitrack DAWs allow an editor to slide one or more audio tracks along a timeline in juxtaposition to other tracks. This attribute allows for more accurate synchronization among individually recorded audio tracks. For example, sound effects and voice tracks in radio commercials may be precisely positioned to produce the greatest impact or the most satisfying sound. Of course DAWs also allow a user to adjust sound levels of individual tracks and to place or to pan signals within the stereo field. With the appropriate plug-ins, DAWs permit an operator to add effects, to manipulate the pitch of signals, and to adjust the duration of tracks by stretching or compressing them. In short, a finished audio production can be manipulated in the editing process in ways that would have amazed audio professionals twenty years ago.

One thing has not changed: Creative productions still require imagination and a grasp of audio aesthetics on the part of producers. DAWs are powerful tools, but they remain merely tools. In the hands of producers who combine an understanding of the use of these tools with a knowledge of communication processes and the aesthetics of production, DAWs can create true audio art.

FOUNDATIONS OF EDITING: ANALOG TAPE

The typical digital audio workstation builds on, and largely emulates, familiar aspects of analog editing through its user interface. For this reason, and the fact that analog tape is still used in some settings, it is helpful to review the basic techniques of editing analog audio tape.

As discussed in Chapter 6, Recording, Storing, and Playback of Sound, when analog tape moves along its path from feed reel to take-up reel on a professional three-head audio tape deck, it first passes the erase head, then the record head, and finally the playback head. Figure 8.1 shows the head assembly on a typical reel-to-reel machine. The gap in the playback head is the point at which the magnetic patterns embedded in the audio tape are converted to corresponding electronic variations that, in turn, are converted into the sound heard by the editor when selecting edit points.

In linear tape-based editing, locating the general section of a tape to be edited requires playing or fast-forwarding through the reel of tape to the approximate lo-

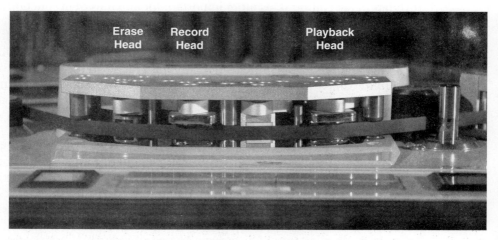

FIGURE 8.1   On this typical reel-to-reel head assembly loaded with tape, edit points are found at the center of the playback head where its gap is located. The playback head gap can be seen easily with magnification.

cation where the recorded material is to be edited. In long productions, this can be extremely time-consuming, requiring many stop-start repetitions. Identifying edit points on tape recorded at faster speeds will be easier than on tape recorded at slower speeds because of the greater length of tape used at higher speeds. In addition, marking edit points will be more precise at higher speeds.

Once the approximate location on the tape is found, the process of marking the spot to cut the tape is next. On most professional tape machines an edit button can be used to defeat its tape lifters, which allows the sound to be heard as the tape is slowly pulled past the playback head by hand. The editor can then hear the exact point where the cut is to be made. While pulling the tape back and forth across the head, the tape's tension must be carefully maintained, so that it neither stretches nor drops away from the heads.

For speech, the tape must be stopped as close as possible to the beginning of the word or phrase to be deleted without cutting off any of the beginning of the first word. For example, in order to remove the word "uh" from the recorded statement, "We, uh, are ready," the tape should be cued to immediately before the beginning of the utterance "uh." That precise point is marked using a grease pencil or china marker by making a small, thin mark on the backing of the audiotape at a point corresponding to the gap in the playback head. Next the tape must be cued to the beginning of the next word to be retained in the final edited version. In this example, "We, uh, are ready," the tape is next cued to just before the beginning of the word "are," where another thin mark on the tape at a point corresponding to the gap in the playback head is made. Figure 8.2 illustrates how the tape would be marked in this example, if one could actually visualize the content of audiotape. The vertical marks before and after "uh" represent the grease pencil's edit points.

The tape is pulled away from the head assembly and placed in the groove of an editing block with the edit mark visible. Most editing blocks have both a diagonal and a vertical slot to guide the blade to cut the tape. The diagonal groove is

FIGURE 8.2   Tape recording marked for editing.

FIGURE 8.3   Cut tape in the splicing block.

used to guide a one-sided, sharp razor blade in cutting the tape into two sections at the first of the two edit marks. Generally, the diagonal slot is used because once the ends of the tape are rejoined, the splice tends to be more secure and passes more easily over the tape heads during playback. Next, the second edit mark is cut in the edit block using the razor in the same manner. Figure 8.3 shows the tape in the splicing block after being cut apart.

To complete the edit, the two loose ends of the master tape, in this example the end with "We" and the end with "are," are placed together in the editing block groove, as illustrated in Figure 8.4. The ends of the audio tape are positioned so that they meet snugly and a short length of splicing tape is used to join the two ends of the master tape.

Finally, the tape is rethreaded onto the deck and played to confirm that the edit is not obvious. If the rhythm of speech does not seem natural, the edit must be redone, either to allow more space between words or to reduce the length of pauses. It should be evident that proper tape editing requires skill, dexterity, patience, and experience.

EDITING WITH DIGITAL AUDIO WORKSTATIONS

An important difference between editing on tape and editing on a workstation is the relative ease of manipulating nonlinear recordings in electronic form. Modern digital audio workstations make it possible not only to remove, or cut, undesired audio elements from the production and easily rearrange recorded sequences but also to restore removed material and to adjust the material for pacing and tempo. The DAW can instantly access any point in the program material merely by pointing the cursor to a desired location on the timeline—no more fast-forwarding and

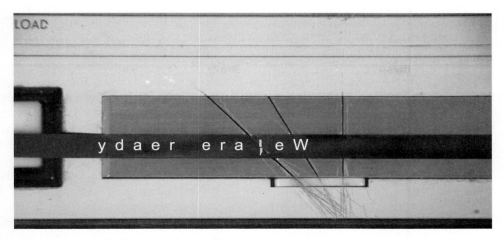

FIGURE 8.4   Edited tape ready to splice.

playing to locate the right spot on a tape. Use of a workstation's "undo" function can quickly restore any audio material removed in error. And, as described earlier, the DAW allows an editor to add a variety of special effects such as equalization, reverberation, echo, and flanging. The software may also allow the editor to shift the pitch, reverse the signal as if playing backward, and invert its phase.

In addition to editing functions, a digital workstation can serve as a recorder capable of capturing analog or digital audio signals on a real-time basis. Many software systems allow workstations to act as multitrack recorders, capturing eight, sixteen, or possibly more audio inputs simultaneously. The number of tracks a system can handle is usually dependent on the software employed and the speed of the DAW's processor.

ADJUSTING LEVELS AND EDITING

Most DAWs allow the editor to adjust the level of any portion or all the audio tracks and to pan individual signals along each timeline. When used in this way, the workstation behaves as a mixer. The advantage of a DAW when used as a mixer is that gain need not be adjusted in real time, but can be preprogrammed by the operator. Levels can be tested and adjustments made whenever needed. Moreover, as noted in the preceding chapter, more powerful DAWs allow the user to access a variety of software plug-ins that make it possible to manipulate sound in additional ways. These and the built-in processing capability of most systems permit users to remove various kinds of noise such as clicks, pops, and hiss; and to stretch or compress the running time of productions.

The power of DAWs' editing capability may be best understood, however, by considering how the software actually handles the audio material. An important concept of a DAW's operation is that a change made to the sequencing of the audio material is only a change in playback sequence. This means that if a user cuts an audio segment, it is not truly removed—the edited segment is just not played back. The original file remains unchanged, or at least unchanged until the changes have been saved. Because the software manages sets of digital information, and not a linear waveform, the editor is required to locate and mark stored data for playback. When the instruction to cut a passage is registered, the software merely skips that data on playback. Some software systems permit the user to undo multiple levels of edits, but others have only a restricted ability to reverse edits. For this reason, and because it is always wise to back up frequently, a good editing practice is to retain the original file, and to save edited versions of the project under different file names as work proceeds using the DAW's "save as" function.

An operator may select a portion of the audio file—that is, the digitized waveform—then copy it and paste it elsewhere in the production. It should be emphasized that the visual representation of the copied waveform is for the benefit of the user, not the actual signal recorded by the workstation. The editing software alters the playback sequence of the chosen data rather than actually moving data. The copied selection may be pasted into the timeline as many times as desired. In most workstations, the data set will not really be duplicated (which would increase the file size dramatically), but the editing software will be instructed instead to play back the copied section multiple times. In systems using this approach, the number of instructions will increase, but the data file size will not. It is this capability that makes this kind of virtual editing "nondestructive." Source files are unchanged by the software, but sets of instructions guide the editing process.

Of course, once editing is completed and a file is saved or converted to another format for storage, export, or transport, it will again become a linear data stream. This stream will be stored in a new audio file, based on edit instructions. For instance, as an edited audio file is converted to a compact audio disk, any segments that have been copied and pasted will actually be duplicated on the CD. The audio remains in the digital realm, but it will be a new linear stream of audio data.

The basic process of editing using a digital audio workstation loosely parallels editing by means of a word processor—at the same time borrowing the look and to some extent the feel of analog tape editing. A workstation's visual interface makes achieving undetectable edits much easier than the somewhat cumbersome methods used in editing analog tape. Digital audio editors make it possible for producers to highlight a segment to be removed and then to select a cut function from an on-screen menu. With some editors, this process is as simple as placing the cursor at the start of the region to be cut, clicking and dragging it over the region, and then selecting "cut." Other editors may require "in" and "out" points to be selected individually, followed by a "cut" command. In most editors, the cut command can also be executed from the computer keyboard, often by the "delete" key or by some other designated key or key combination. Other commands such as "copy" and "paste" can also be selected from the menu or from a designated key.

When editing speech, it is important to make edit selections on the on-screen waveform display with care, taking pains not to clip word beginnings or endings. It is also important to maintain proper spacing between words, so that the edited speech rhythm seems natural. Normal speech patterns usually contain slight pauses between words. Words that lack this spacing run together unnaturally and will alert listeners to editing of the speech. When editing music, it is usually preferable to mark edit points on the waveform display at the beginning of musical measures, on the beat. If the intent is to merely shorten a piece of music, the operator should strive for a natural result by maintaining its rhythm—in other words, no abrupt changes in tempo or beat count. Familiarity with musical structure can help an editor judge the musicality of edits. Some editing systems allow a user to select the timeline scale, usually offering the options of displaying running time in hours, minutes, and seconds; in frames in motion picture time-code; or in beats per minute.

To minimize the introduction of annoying pops, clicks, or other artifacts at edit points, cuts should be made at the point where the audio signal's waveform is at or near zero amplitude. Using the waveform view, this can be readily seen by finding the spot at which the waveform crosses, or nearly crosses, the zero amplitude line. Figure 8.5 illustrates a waveform view in an audio editing application. Note the point at which the waveform's amplitude falls to zero. These are the points at which edits can be made with less risk of creating noises or calling attention to the edit; they represent the momentary points of silence between words or beats. Figure 8.6 shows a section of silence between phrases highlighted for removal, and Figure 8.7 illustrates the waveform after removal of the highlighted section. Whether editing speech or music, audio producers take pride in making their edited works sound as though they had never been altered. The basic rule is this: avoid drawing attention to edits; try to make unobtrusive edits so that a listener's focus will be on the content, not the edits.

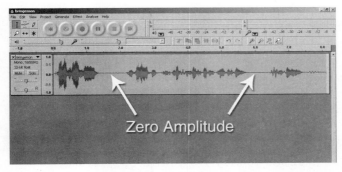

FIGURE 8.5   This waveform view in an Audacity editing workstation shows an audio signal as it drops to zero amplitude.

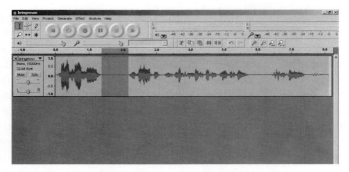

FIGURE 8.6   Silence highlighted for editing.

Some editors provide a **"scrubbing"** function. Scrubbing simulates the action of tape reels rocking back and forth moving tape across the playback head of a reel-to-reel machine. Scrubbing allows the operator to hear the signal as the cursor is moved along the waveform or timeline. Dragging the cursor quickly along the waveform produces a sound like that of fast-forwarding or rewinding tape in contact with a playback head. Using scrubbing techniques, an editor can audibly locate momentary pauses between words or musical beats in order to mark edit points. Although less exact than using a visual waveform to locate zero-amplitude points, scrubbing can nevertheless provide a better feel for the pacing and rhythm of edits. For producers who are accustomed to tape editing, the scrub function may seem more natural than the use of on-screen displays and may help them edit in a way that is more like conventional tape editing.

Many editing applications offer two on-screen methods for adjusting levels. One is the familiar fader, which is often graphically represented on-screen. Placing the cursor over the fader, clicking the mouse, and moving the on-screen fader will change the output level of its associated signal. In most editors, level changes made during real-time playback of the files are stored and recalled on the next

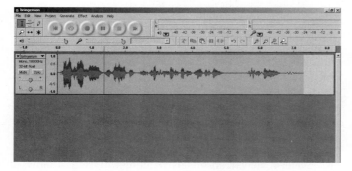

FIGURE 8.7   Waveform after removal of highlighted portion.

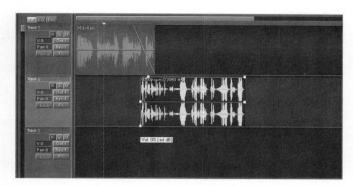

FIGURE 8.8   Waveform marked for envelope fade.

playback. This integrated automatic mixing function is another powerful feature of digital workstations. For example, in a multitrack editor, the level of an individual track may be faded down at a certain point in the timeline. That fade will be recalled during each playback, unless it is subsequently undone.

A second mechanism for altering levels in many editing software suites is the visual envelope marker. As explained in an earlier chapter, the term *envelope* is used to describe the attack, sustain, release, and decay characteristics of a sound. This term was used previously to describe a discrete auditory event such as the nature of a drumbeat: fast attack, short sustain, and rapid release and decay. In audio editors, the envelope may be extended to encompass the entire program or timeline. The software used by workstations often allows the user to place control points along the envelope that can be repositioned to adjust playback level. Envelope adjustments in some editing systems are made more intuitive by presenting a visual cue to relative levels of tracks at a glance. In Figure 8.8, this can be seen in an audio envelope that has been set to fade out at its end.

Audio editing software packages all share a similar on-screen appearance. This commonality allows an operator to make the transition from one editing system to another without difficulty. While each software editor may possess certain special features and capabilities, they will all include a waveform display or timeline, shuttle controls, audio level controls or mixer, and menu-driven effects processors. To take full advantage of the capabilities of any workstation suite will require practice, but basic functionality can be immediately exploited—or at least quickly learned—by most experienced audio engineers. Note the similarities in Figures 8.9 through 8.14, which show the on-screen appearance of five popular audio workstation systems. Figure 8.9 depicts Digidesign's Pro Tools, Figure 8.10 is a screen shot from Adobe's Audition, Figure 8.11 is from Sony's Sound Forge, Figure 8.12 shows the on-screen appearance of Audacity, Figure 8.13 illustrates Digital Audio Restoration Technology's DART, and Figure 8.14 presents Apple's GarageBand.

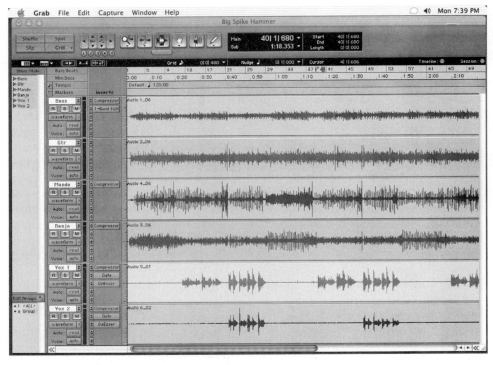

FIGURE 8.9 Digidesign Pro Tools.

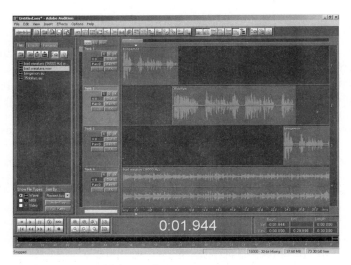

FIGURE 8.10 Adobe Audition.

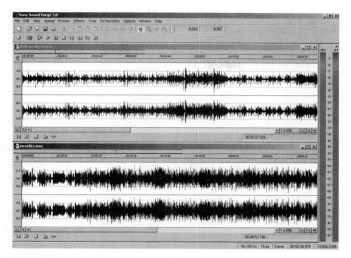

FIGURE 8.11   Sony Sound Forge.

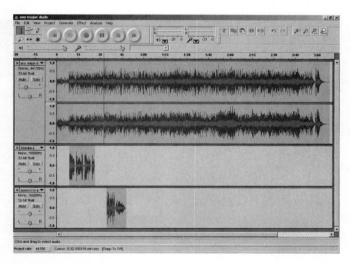

FIGURE 8.12   Audacity.

FIGURE 8.13 Digital Audio Restoration Technology DART.

FIGURE 8.14 Apple GarageBand.

CONTROL SURFACES IN EDITING

As mentioned elsewhere, many audio professionals find it preferable to employ a tactile hardware interface or control surface with the DAW, rather than conventional computer input devices. Volume control using a real fader is considered easier and more intuitive by some operators. For example, many producers believe that executing a long, slow, smooth fade will be easier with a fader than with a mouse or track ball. Hardware interfaces may provide only for the control of on-screen functions, but some units also contain microphone preamplifiers and connectors for line-level inputs. The interface shown in Figure 8.15 is purely a remote control that allows an operator to adjust levels and other software parameters, but it lacks any ability to handle audio signals. A more elaborate control surface is provided by Digidesign's model Digi 002, shown in Figure 8.16 (page 122).

FIGURE 8.15   This tactile interface is purely a control interface and does not provide any audio processing functions.

FIGURE 8.16 The Digidesign Digi 002 control surface provides both a control interface and an audio interface. At the rear of the unit are computer connectors as well as audio inputs and outputs.

CHAPTER NINE

AUDIO PROCESSORS AND PROCESSING

This chapter considers the range of devices and applications that can be used to alter sound qualities of audio signals. Whether controlling volume levels or equalization, or artificially adding ambience through delay and reverberation, **processors** are indispensable tools for creative sound recordists and recording engineers.

AMPLIFICATION AND LEVEL CONTROL

The simplest kind of electronic processing is level amplification. Increased audio levels can only be accomplished by means of active components, that is, components that produce voltage or power **gain**. Active devices require electronic power to produce amplification. Active devices include vacuum tubes, transistors, and, more commonly today, integrated circuits—small devices that contain multiple transistors and other components within a single package. Most audio equipment includes at least one **gain stage**. Due to signal reductions, or attenuation, introduced in the electronic circuits of processors, gain must be added to offset losses. Pieces of professional equipment in contemporary studios are normally designed to maintain **unity gain**. Unity gain merely means that output levels are kept equal to input levels. Hence, if an input signal is equal to a peak of 1 volt (V), then the unity gain output will be 1 V at its peak. This is also known as a 1:1 gain ratio. Circuitry in a unity gain device will compensate for its inherent signal losses by adding just enough gain so that the output levels are not lower than the input levels.

As noted earlier, mixing consoles incorporate preamplifiers to raise low-input audio levels received from microphones and to maintain levels input from other audio sources. Microphones normally produce a sound output that requires something on the order of 40 dB or more of amplification to raise the audio signal to line level. Controls on a mixing console can be set for unity gain, so that output signals will be maintained at line level.

Power amplifiers have a number of gain stages that together successively boost **line level** input signals to output levels high enough to move the heavy voice coils of large loudspeakers. Two or three different line level standards are employed in professional settings, but the traditional standard is equivalent to one **milliwatt** in a low impedance circuit. (A milliwatt is one-thousandth of a watt.) In contrast, power amplifiers are typically rated in watts, a **watt** being a measure of the "rate of doing work" by the device. Wattage is sometimes also described as a measure of power "consumed" by a device. Most people are familiar with the term *watt* from everyday devices such as light bulbs and hair dryers, and from the way electrical service is billed in units of watts each month.

Watts are the units used in audio to measure the power of a power amplifier, or the amount of energy it can supply at its output to a loudspeaker system. To increase the apparent level of an audio signal by 3 decibels (the smallest increase most people can easily hear) requires an amplifier to double power. Each time audio level is increased by 3 dB, an amplifier's power output will double. So, for example, to achieve a 3 dB increase in output of a 20 watt (W) amplifier, an increase to 40 would be needed. Amplifiers range from low-power phonograph and microphone preamplifiers that boost a signal to only a few milliwatts to large commercial-power amplifiers that can deliver thousands of watts. The amount of gain that a single circuit stage can provide is limited, so that amplifiers usually must be constructed of several gain stages, each one of which progressively boosts signals to higher power levels.

Compared with amplification, decreasing audio levels is simple and may only involve the use of an **attenuator** such as a potentiometer, also called a fader. Potentiometers, or pots, are nothing more than an extended pathway for the current to follow, a path with increased opposition to current flow. As electrical current flows through a conductor, it will gradually exhibit some voltage drop. The higher the resistance of the path through which a signal travels the more it will be attenuated, or reduced in level due to opposition in the pathway. Pots may also be accurately described as **variable resistors**, because they allow the operator to vary the amount of resistance added to the signal pathway.

Conventional pots are constructed of extended lengths of high-resistance wire, wound in either a circular or linear form. However, potentiometers used today in audio equipment often make use of a tough, highly resistive material rather than a wire. In a potentiometer, the audio signal flows into the resistive material, and then exits through a wiper attached to the pot's shaft or fader control. The construction of a potentiometer can be seen in Figure 9.1. Both linear and rotary pots are shown in this illustration, and it can be observed that the farther the wiper is located from the current's input, the greater will be the attenuation.

If a fixed amount of level reduction is required, it may be possible to use a pad. Pads are passive devices that merely insert a predetermined amount of resistance into a signal pathway, introducing a comparable reduction in output levels. Pads are categorized according the amount of reduction measured in decibels that they offer. Pads are often integrated into mixing consoles, high-quality condenser microphones, and other devices to help cope with output levels so great that sig-

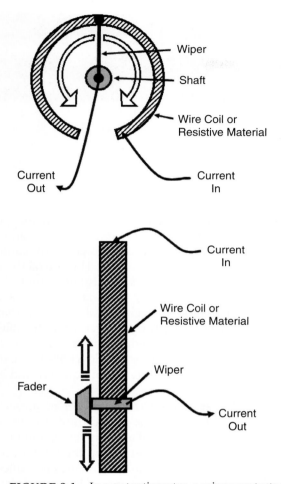

FIGURE 9.1   In a potentiometer, a wiper contacts a strip of resistive material arranged either in a circular or straight flat shape. As the wiper slides farther away from the input, the potentiometer's resistance increases.

nal clipping in attached equipment might occur. Pads are also available as standalone devices that are designed to be inserted inline, or plugged directly into cabling of microphones or instrument lines. As can be seen in Figure 9.2, inline pads look like connectors, and matching cable connectors can be inserted into one or both ends. Typical ratings for pads might be 10 or 20 dB, with some units allowing the user to choose among two or more reduction levels by means of a switch.

FIGURE 9.2 An inline pad.

DYNAMIC PROCESSORS

A **limiter**, a **compressor**, and a **noise gate** are classified as dynamic processors, and all three provide some form of automated level control when parameters are preset by the user. Dynamic processors alter an audio signal's dynamic range— that is, the range in level between soft and loud sounds. This is exactly what happens when a mixer operator strives to achieve a consistent output by bringing level faders down when audio levels rise and by pushing faders up when levels drop. The problem with a manual method of level control like this is that changes in source levels are likely to occur faster than the operator will be able to react. Once a user has set operating parameters, gain reduction circuits in these dynamic processors assume a portion of control over the signal's level. By doing this in real time, they shrink the sound's dynamic range much more quickly and accurately than a human operator could ever achieve.

Limiters, sometimes called peak limiters, allow a user to preset a certain maximum output level, beyond which output signals cannot rise. For example, the user may determine that zero on the scale of a VU meter should be the maximum output signal allowed by a limiter. After setting up the limiter accordingly, any signal that exceeds zero VU will be rapidly and automatically reduced to that level. Limiters provide the user with threshold and output-level controls, but often little else. The **threshold** is the level at which the limiter begins to decrease its signal's amplitude. When a signal rises above the threshold level, the limiter decreases its gain to prevent its output from exceeding the preset level. Limiters are usually characterized as having fast release times. **Release time** refers to the amount of time that lapses during which a limiter maintains signal reduction after input levels drop below the threshold. Limiters are processors commonly used in broadcasting, because radio and television transmitters must be protected from excessive signal levels that would cause **overmodulation**. Overmodulation is prohibited by regulatory agencies because it causes severe distortion of the broadcast signal and may produce interference to other stations. Audio engineers sometimes criticize limiters for generating a harsh sound; this results purely from the operation of the gain reduction circuits. Like other kinds of processors, limiters must be used with care; application of minimal processing is usually the best strategy to avoid unwanted by-products.

Compressors resemble limiters in certain ways, but they offer technicians greater control over levels. In addition to limiting maximum signal levels, compressors can also reduce the level of a signal throughout its dynamic range. They also permit users to vary the threshold level and the **ratio** of the reduction, as well as the attack and release times. As with limiters, the threshold setting determines the signal level at which a compressor begins its gain reduction.

Compressors' ratio control enables users to set the gain reduction rate that occurs once input signals have exceeded the threshold setting. If a compressor is adjusted so that it will compress signals at a 2:1 ratio, it will reduce output audio 1 dB for every 2 dB that the input signal rises above the threshold. A compression ratio of about 8:1 or 10:1 is considered severe, and will eliminate much of the natural dynamics of a musical performance. Consequently, compression ratios of 2:1 or 3:1 are usually preferred in order to retain transients and intentional changes in sound level.

Most compressors allow users to determine how quickly gain reduction engages once an input signal exceeds the threshold. This characteristic is called **attack time**. Attack settings should be set to approximately match the envelope (attack, sustain, delay, release) of an incoming signal. Short attack settings are needed in order for a compressor to respond to signals having sharp attack envelopes, such as percussion instruments. In compressors, as in limiters, the length of time a compressor continues to reduce gain once triggered into action is called the release. Long-release settings maintain gain reduction between short repetitive sounds, such as drumbeats, while fast release times allow a compressor's gain to recover between beats.

Noise gates present an all-or-nothing sort of gain control. As the name implies, a gate is a device that can either be open to pass signals or closed to block them. A noise gate will prevent any signal from passing through when the input signal falls below its threshold level, but when the threshold is surpassed signals will pass through unaffected. The gate's threshold should be adjusted by the user to open only when a desired audio signal is present. For example, when a recording studio engineer places a gated microphone near the head of a snare drum, the threshold should be set high enough so that the gate will block sounds of nearby instruments and open only when the microphone picks up sounds of a direct solid hit on the snare drumhead. This technique not only effectively mutes the sounds of sticks hitting other instruments but also blocks sounds of incidental rim strikes as well as spurious rattles and mechanical noises from the drum kit. By using a noise gate this way, producers can achieve a clean track containing only sound picked up from the snare.

FREQUENCY PROCESSORS

Control of audio signal levels across a range of frequencies can be achieved through the use of either filters or equalizers. At the most rudimentary level of control is the **shelving** filter, which passes a large band of frequencies but cuts large bands at either end of the frequency spectrum. The **high-pass** shelving **filter** allows high frequencies to pass while reducing audio output of signals below a selected cutoff frequency. Conversely, the **low-pass filter** reduces all high frequencies but allows low frequencies to pass unaltered.

Conventionally, a **filter** is a passive device, meaning that it adds no gain and cannot boost an audio signal. Therefore, filters can only cut or deemphasize portions of audio signals. One filter type is known as the LC filter, so called because its filter circuits are constructed from inductors (L) and capacitors (C). When joined together in a circuit, inductors and capacitors become electronically resonant. Resonant circuits are able to reduce or pass signals at a fixed frequency or band of frequencies. Hence, LC filters are termed *resonant*, or *resonators*.

Popular filter circuits include the Butterworth and Chebyshev types, both named for the persons who developed the designs. The Butterworth filter is widely used because it exhibits little irregularity in response across the band of frequencies passing through it. Other filter designs such as the Bessel and Gaussian can be found in audio devices, too. One common application for filters is in speaker crossover networks. When used in these networks, one filter removes the high frequencies being passed to the woofer and another filter screens out the low frequencies passed to the tweeter.

Although filters are usually passive devices, some contemporary filter units are designed as active devices. Active filters use amplifiers to generate the effect of circuit resonance. They are especially useful in cases where use of an LC filter would require a large, cumbersome inductor.

Shelving filters are used in removing sounds from recordings that predominate at the upper or lower ends of the audio frequency spectrum. Examples include low-frequency rumble from air-conditioning equipment and high-frequency tape hiss produced by an old recording. The shelving filter's main drawback is its broad, nonselective effect on frequency response. The filter will remove all sounds of the affected frequency range, including not only offensive sounds but any others that fall within the spectrum being reduced. The frequency characteristics of a typical high-pass shelving filter is shown graphically in Figure 9.3.

A **notch filter** provides users the ability to reduce a selectable narrow band of frequencies. The notch filter gets its name from the visual appearance of a frequency response curve having a flat profile except for a single narrow deemphasis band resembling a notch. An example is shown in Figure 9.3. Notch filters can be useful for removing a troublesome hum or deemphasizing a narrow band of frequencies. A **comb filter** provides a series of very narrow bandwidth notch filters. When several frequencies are notched out of the frequency response, the resulting curve resembles the outline of a hair comb, giving the filter its name. Notch filter circuits can be integrated in the design of other processors. Much of the effect created by processors such as **flangers** and chorusers is the result of shifting notches in the audio spectrum.

Equalizers vary in their design and complexity, from the simple **sweep** equalizer and shelving equalizer to the precise parametric and graphic equalizers. (Colloquially, producers and engineers refer to an equalizer as "EQ.") Unlike fil-

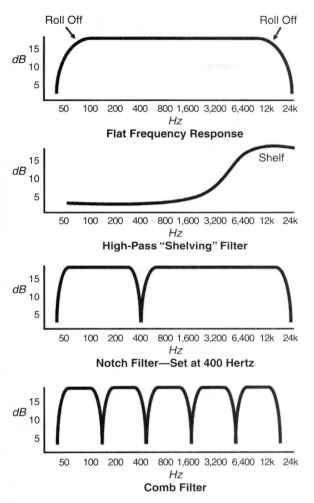

FIGURE 9.3   Four different filter types and their frequency response curves.

ters, an **equalizer** employs active devices that either emphasize (boost) selected frequencies or deemphasize (cut) them. To boost the signal at selected frequencies, an amplifier incorporates resonant circuits that shape its response characteristics. Variable resistors are also built into the circuit in order to make the boost or cut in frequency gain adjustable by the operator.

The sweep EQ is common on inexpensive consumer equipment. It normally takes the form of a simple frequency control that emphasizes either bass, mid-range, or high frequencies. Rotating its control allows the user to "sweep" the frequency spectrum, applying boost as it sweeps. Figure 9.4 presents a graphic representation of the sweep equalizer.

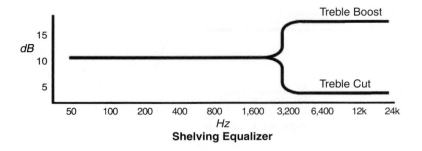

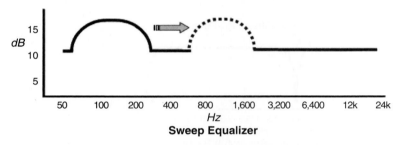

FIGURE 9.4   Two types of equalizers and their corresponding frequency response curves.

The shelving equalizer, like the shelving filter, affects frequencies at the upper or lower end of the audio spectrum. However, the shelving EQ is able not only to reduce a band of frequencies, it is capable of boosting them as well. The shelving EQ normally incorporates multiple bands (e.g., bass and treble, or bass, mid-range, and treble). Boosting the bass control, for example, will increase a band of low frequencies; cutting the bass control will decrease that same band of low frequencies. The range of the frequency bands affected by a shelving equalizer is fixed and cannot be modified. Like the sweep equalizer, the shelving equalizer is more often seen on inexpensive nonprofessional equipment.

A **graphic equalizer** is designed to present users with a visual representation of the equalization curve applied to audio passing through it. In this type of equalizer, narrow bands of frequencies may be adjusted via sliding faders, and it is the position of the fader controls that visibly indicates the EQ curve. If a U-shaped curve were shown by faders, it would indicate that the bass and treble frequencies were boosted while the mid-ranges were deemphasized. Graphic equalizers vary in the width of frequencies affected by each fader control. The narrower each control's bandwidth, the larger the number of faders needed to span the whole audio spectrum. Most graphic equalizers contain ten to thirty bands of controls. These

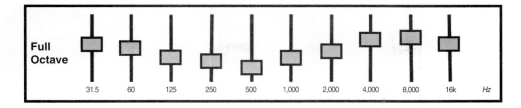

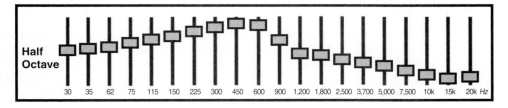

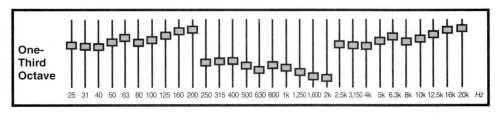

FIGURE 9.5   Shown at the top is a full-octave graphic equalizer, and below it are one-half-octave and one-third-octave graphic equalizers.

kinds of equalizers have a number of specific applications, for instance they are often found in the signal path of live sound reinforcement onstage monitor systems.

The bandwidth affected by faders in graphic equalizers is described by the intervals between center frequencies of the preset bands they represent. Common examples are *full-octave, half-octave,* or *one-third-octave* equalizers. A full-octave equalizer has ten bands, and the center frequencies are each one musical octave apart. A twenty-band equalizer has center frequencies that are separated by one-half octave, and so on. Figure 9.5 shows controls of a graphic equalizer and Figure 9.6 presents a photograph of a unit. The precision of a one-third-octave equalizer makes it possible for a user to "notch out" a single annoying resonant frequency without seriously affecting overall response. Conventionally, graphic equalizers are manufactured to match bands of frequencies defined by the International Standards Organization, referred to as ISO Center Frequencies. With few exceptions, graphic equalizers are manufactured as *outboard* devices, meaning that they are separate pieces of equipment not integrated into devices having other functions.

A greater precision in control over frequency characteristics is afforded by a **parametric equalizer**. In these equalizers there are adjustments available for

FIGURE 9.6   Front panel of a one-third-octave graphic equalizer.

three distinct parameters—or characteristics—of audio signals. These can be seen in Figure 9.7. First, parametric equalizers allow a user to select the center frequency affected by processing. Second, selected frequencies may be boosted or cut by a user-defined amount. And third, a user may also adjust the bandwidth of frequencies affected. A user may choose to alter a narrow band of frequencies or a broad band, depending on the need. Some manufacturers call this feature variable "Q" to distinguish equalizers like this from other designs that do not permit an operator to alter the bandwidth of frequencies boosted or cut. Most parametric equalizers provide multiple sections to allow simultaneous treatment of the three parameters in two or more portions of the audio spectrum. Parametric equalizers are frequently integrated into studio recording consoles within individual channel strips. A photograph of a five-section parametric equalizer can be seen in Figure 9.8.

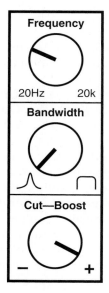

FIGURE 9.7   The controls of a parametric equalizer are used to set its three primary audio parameters.

FIGURE 9.8   Front panel of a five-section parametric equalizer.

TEMPORAL PROCESSING

When processing alters timing relationships within audio signals, the result is termed a **temporal** effect. Temporal processing simulates effects resulting from the lapse of time that occurs when a sound wave travels through space. In audio, the word **delay** is a general term that describes a time offset between a sound and a repeat of that sound, normally at a reduced level. In many cases it appears as a short-interval echo. In nature, delay results from the reflection of a sound wave from nearby surfaces, but it can be created artificially by a number of different electronic means.

Among temporal effects are ones described as echo, delay, and reverberation. These effects are complex and the aural results are highly variable. A common use for these effects is to add ambience to recordings. For example, temporal effects are often employed to imitate sonic spaces within which music performances take place. Ambience is the quality that arises from the presence of slight background sounds that give a natural impression to recordings. A barely detectable quality of reverberation within a room is a major component of ambience.

An echo is customarily defined as the effect caused by the single discrete reflection of a sound wave. An echo's effect is related to the time that passes as sound travels to a distant surface and reflects back to the listener. Traveling at 1,130 feet per second, a sound wave will be heard as a distinct and audible echo if it completes a round trip in as little as a few hundred feet. Echoes may be repeated, but they will still be heard as separate and distinct sounds if they are separated by intervals of about one-tenth of a second or longer.

When sound travels within an enclosed space such as a room, waves will reflect from the walls, provided the surfaces do not absorb them to any significant degree. Reflection is efficient from hard flat surfaces, but will be dampened by soft, porous, and irregularly shaped surfaces. As sound continues to bounce within an enclosed space, waves will be reflected again and again, resulting in the effect called reverberation. Reverberation is thus created when sound waves reflect from multiple hard surfaces, producing a sequence of sounds that arrive at the listener's ears delayed by less than one-tenth of a second. Reverberation is a more complicated temporal effect than echo because it results from a complex combination of reflections arriving at the listener's ears in such rapid succession that individual reflections cannot be individually discerned. Separate reflections seem to be blended or blurred together. The short separations between reflections make them appear to be a single sound. This sound can be sustained for quite a long time in the right acoustic environment. Sounds in large open rooms with hard walls, such as cathedral interiors, can continue to reverberate for several seconds. Engineers measure reverberation in the time it takes for the level of reverberating sound to decay by 60 dB, or RT_{60}.

When sounds reverberate, a listener's brain is able to define the approximate acoustic space within which an auditory event is taking place. The length of time required for a reverberant sound to decay and the tonal characteristics imposed on reflected sounds suggest the spatial dimensions of a room. As reflections continue to bounce from surface to surface, their high-frequency content tends to diminish because these frequencies are more easily absorbed by wall surfaces. So, while the first or early reflections may exhibit full-frequency response, as the reverb time extends, high frequencies will be lost. Most enclosed environments exhibit some degree of reverberation introduced by the plaster walls and ceilings used in typical human living spaces.

Beginning in infancy, a listener's brain learns to assess and approximately define enclosed spaces. This can be achieved with remarkable accuracy based solely on auditory cues. Without being conscious of how it is accomplished, sound reverberation enables the listener to visualize a surrounding sound space. Because human environments normally exhibit sound delay effects, the addition of reverberation to a recording lends it a familiar and natural quality. In early days, most recording studios were intentionally designed to be totally nonreverberant, to be acoustically dead. To achieve a natural sound in recordings made in these studios, it was necessary to add synthetic reverberation during the recording process.

One type of delay effect can be simulated by the highly reverberant sound of a tiled restroom. This effect became fashionable in the 1950s, when it was associated with popular rock-and-roll and rock-a-billy records of the day. Musical lore has it that vocal groups of the 1950s would rehearse in public restrooms, bus stations, or subways to take advantage of the ambience that these acoustically live environments added to their performances. The small hard-surfaced spaces helped performers hear each other and blend their voices more pleasingly. Additional discussion of room reverberation can be found in Chapter 15, Multitrack Recording.

Methods for producing reverberation evolved over time from acoustic, to electromechanical, to analog, to digital and, most recently, to "virtual" or software-based techniques. Reverberation chambers were once entirely acoustic; they were created by placing a speaker and a microphone in a room having hard-surfaced walls, floor, and ceiling. Because of the space and expense required to create an acoustic reverberation chamber, recording studios sought less costly and more compact methods for creating sound delay effects.

Early echo techniques sometimes employed tape-based methods. These systems relied on an audio signal sent to a tape machine's record head and played back from its reproduce head as tape moved through the mechanism. If the tape moved at 7.5 inches per second, and the two heads were separated by 7.5 inches, a signal would be played back one-second after being recorded. This created a one-second delay, or echo. Changing either the tape speed or the distance between record and reproduce heads would alter the echo's delay interval.

Early echo devices, such as the Maestro Echoplex, were constructed on this model. The Echoplex design was nothing more than a tape recorder that made use of an endless loop of tape and a moveable playback head. Figure 9.9 shows construction details of a device like this. Shifting the playback head farther away from the record head extended the time that elapsed before an echo would repeat. Moving the playback head closer to the record head shortened the echo's delay. A **feedback** control permitted the operator to adjust the amount of level decay in successive echoes.

Electromechanical techniques were also developed to simulate temporal effects. These were based on diverting an analog audio signal through some sort of extended mechanical pathway. Sound waves can travel through solid materials such as metal, but they do so much more slowly than through the air. By transferring sound to metal springs, steel plates, or gold foil, it is possible to mimic acoustic reverberation within enclosed spaces.

Two electromechanical methods that gained the greatest popularity were the **spring reverb** and the **plate reverb**. Spring reverberation devices are simple and inexpensive to manufacture. Although stand-alone spring reverb units can be purchased, they are more frequently seen as accessories integrated into guitar amplifiers or public-address amplifiers. To achieve a reverberation effect, a spring is stretched taut within a soundproofed enclosure. At one end of the spring a transducer, essentially a loudspeaker driver, is mechanically coupled to the spring. When the transducer applies an audio signal, sound waves travel along the spring as it vibrates. To capture the reverberated signal, multiple transducers somewhat like contact microphones are attached at points along the spring. As the spring vibrates, these transducers convert the physical flexing of the spring into an electronic output. Pickup transducers near the sound source simulate short reverberation times, and those farther away simulate longer reverberation times typical of a larger space. Damping controls on some spring-based units inhibit vibrations along the spring, effectively reducing the decay time produced by reverberation. In more sophisticated reverb units, the relative levels output from the pickup

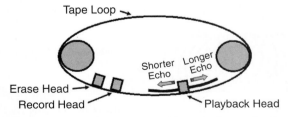

FIGURE 9.9   The operational layout of an Echoplex reverberation device shows how the distance between record and playback heads can be varied to control the delay timing of the reverberant sound.

transducers can be individually adjusted to simulate acoustic environments of various sizes and liveliness (Figure 9.10).

Another electromechanical way of duplicating acoustic reverberation made use of a large steel plate suspended under tension within a sturdy frame. Transducers attached to the plate were employed to induced sound waves in the plate, and in turn, other pickup transducers converted vibrations back into a voltage output in a manner like that described for the spring reverb. Modifications of this scheme have appeared over the years, such as the **foil reverb**, using thinner sheets of metal suspended under tension, many of which were constructed using gold foil. Damping controls on plate and foil reverbs were often added to control reverberation decay time.

Delay-producing techniques advanced a step during the transistor era by means of capacitor-based analog bucket-brigade technology. Analog reverb units were built around devices known as bucket brigade delay (BBD) chips. The BBD drew its name from firefighters' bucket brigades in which buckets of water would be passed from hand to hand. A BBD chip stores electrical charges in a sequence of capacitors—conceptually equivalent to electronic buckets. The charge is passed from bucket to bucket at a rate set by an electronic "clock" circuit producing regularly timed pulses. As a voltage is applied to the first capacitor in a BBD string, it immediately charges. When the clock's pulse is sent, the first capacitor passes its charge to the capacitor next in line, leaving the first empty and ready to store a new voltage input. As clock pulses continue, odd-numbered capacitors dump their voltage into the adjacent, empty, even-numbered capacitors. Then in the next cycle the even-numbered capacitors dump their charge into the odd-numbered capacitors. By this means, voltages are transferred to the BBD output through strings of buckets at a rate set by the clock. Naturally, movement of the signal through the string of capacitors delays it, and to create reverberation, signals will be fed back to the input from selected capacitors along the BBD chain. Varying the

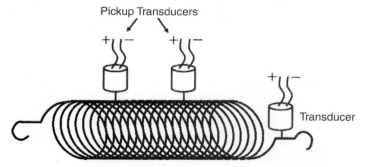

FIGURE 9.10   In a spring reverberation device, sound is introduced by the transducer at the spring end on the right. Sound waves bounce along the spring and are captured by the pickup transducers.

speed of the pulse-generating clock adjusts the length of reverberation delay. Analog reverb devices using bucket-brigade techniques never attained any degree of acceptance in professional equipment. Because they could not easily duplicate the continuously shifting frequency characteristics of successive reflections, their effect was not very convincing. Even in nonprofessional applications these analog reverb devices enjoyed only a brief moment of popularity, one that was cut short by the introduction of much better-sounding digital technologies.

Effects such as a **flanging** or **chorus effect** are created by time delays so short that sound waves combine in opposing phase relationships. These can be considered specialty or novelty effects, since they do not simulate naturally occurring acoustic phenomena. Flanging gets its name from the way that the effect was produced in the days of reel-to-reel tape machines. To produce flanging in the traditional way, two identical tape recordings were played back simultaneously in exact synchrony on two different tape machines. If an engineer pressed a finger lightly on the *flange* of one of the tape reels, that tape slowed down a bit compared with the second tape, and its audio signal was shifted slightly out of phase at certain frequencies. Lightly pressing a finger on the second machine, the producer could then slow that tape to bring signals back into phase. By alternatively slowing first one machine then the other, the two signals could be made to slowly shift in and out of phase. The effect acquired the name *flanging* because it was achieved by the touch of an engineer's fingers on tape flanges. The shift in phase between the two tapes created an extremely sharp comb filter effect and, by careful application of finger pressure, the filters' frequencies could be shifted and controlled. The resulting sound can take on peculiar qualities depending on frequencies affected and the comb filter's depth, at times producing a hollow or ghostly impression.

Recording studio folklore sometimes credits legendary producer Phil Spector with inventing flanging, while other stories attribute the discovery to Beatle John Lennon or Beatles' producer George Martin. These stories are interesting, but they are undoubtedly incorrect because one of the authors learned the technique from other audio engineers in the 1950s, long before either was an active producer.

Generating flanging effects by traditional means is awkward and imprecise, so devices producing the effect electronically have been developed. To electronically produce this effect, an unprocessed audio input is split into two signals, one of which is passed through a variable time-delay circuit before recombination. The time-delay circuit continuously shifts the length of delay so that the two signals go into and out of phase. As this happens, frequencies that are boosted or cut vary along with the delay. When delayed signals are mixed with unprocessed input signals, **doubling** or **slap-back echo** effects can also be created. Devices that produce this effect provide control over the balance between delayed and input signals, the length of time between reflections, and the decay rate of successive reflections.

Contemporary reverb processors offer control over several key signal characteristics. Among the parameters normally available for adjustment are **early reflections**, **predelay**, **reverb time**, **damping**, and equalization. Each one of these provides cues to the brain related to the size, shape, and surface treatments of the space in which sounds are heard. The early reflection parameter controls the length of time before the first reflected sound is heard. The quicker the first early reflection arrives, the shorter is the perceived distance to enclosing walls. The pre-delay parameter affects the length of time before reverberation is initiated. Reverb time and damping affect the length of time reverberation is sustained before dying away. And finally, the equalization control affects the frequency content of reverberated signals.

DIGITAL SIGNAL PROCESSING

Digital signal processing (DSP) is accomplished in a totally different manner from ones used in analog audio processing. In theory at least, DSP has distinct advantages over the techniques described earlier in this chapter, and DSP is now preferred over analog approaches in many, perhaps most, production settings. Digital processing requires that audio be present in digital form, that is, as a series of numbers representing a sound waveform. Each number is a value that equates to an instantaneous voltage of that waveform and is called a sample, as described in Chapter 3, Digital Audio. Using mathematical models, it is possible to modify values of sample numbers so that waveforms are changed in the same way they might be reshaped by conventional analog processing.

The mathematics involved in signal processing is well beyond the scope of this presentation, but the basic processes are not especially difficult to explain. Many data conversion steps in signal processing make use of Fourier transforms that are coded into software routines. Fourier transforms analyze the frequencies contained in a sampled waveform and determine their levels. From these results, computer algorithms can be used to set samples to new values that accomplish desired changes in the audio waveform. Samples can be modified, for example, to boost or cut selected frequencies, to reshape overall audio levels as in a compressor or limiter, or the samples can be partially time-shifted to simulate reverberation effects.

Digital-delay processors achieve their effects using circuits that incorporate delay lines made up of a series of memory locations. When digital data arrives at the digital-delay line, it is held temporarily in the first storage location then after a set length of time it is released and sent onward to a second storage location, somewhat like the bucket-brigade process described earlier. Because of the manner data shifts from one memory address to the next, digital-delay lines are also known as shift registers. Digital audio moves from one location to another in many steps, each step delaying the signal's arrival at the processor output. Some storage locations are called *taps* and they send outputs not only to the next location but also directly to the processor output. This gives the processor an ability to incorporate multiple simultaneous delay times to imitate the sound of enclosed spaces. Audio samples are continually pushed farther along the memory line as the digital stream continues to flow, until ultimately the data arrives at the output.

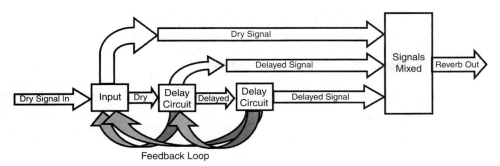

FIGURE 9.11   Signal flow within a digital reverberation device.

This process is illustrated in Figure 9.11. Examples of digital reverberation and digital delay can be heard on the CD-ROM accompanying this text.

Employing a similar processing model, a digital equalizer taps data streams at one of its memory locations, inverts values so that the waveform is out of phase, then returns them to the input stream. Depending on which locations are tapped and the level of inverted signals that are reintroduced, the frequencies affected and amount of equalization can be controlled.

Digital signal processing can be accomplished either in stand-alone pieces of equipment or by software incorporated into DAWs. Stand-alone units are commonly seen as multi-effects processors. These devices incorporate ADC/DAC circuits to convert input analog audio signals, process data in digital form, and then send out the result as an analog signal once again. The Yamaha SPX-90 was a unit that found wide acceptance in the days before digital audio workstations were common. This device was capable of compression, limiting, equalization, and many kinds of reverberation. Later, other manufacturers introduced their own multi-effects processors, including Alesis, Lexicon, and Eventide. Multi-effects units can reproduce practically any effect one can imagine, and they are generally very easy to use. However, as noted elsewhere, most professionals prefer to avoid moving audio in and out of the digital domain in the production chain multiple times.

Most outboard delay devices in use today employ digital processing. Figure 9.12 portrays a typical digital processor. As noted, these devices simulate effects

FIGURE 9.12   DeltaLab digital delay device front panel.

produced by analog acoustic or electromechanical reverb devices but with a much higher sonic quality. With the introduction of large-scale integrated circuits, it became possible for complex digital circuits to be manufactured inexpensively and in smaller sizes. Digital reverb devices are composed of a series of adjustable delay circuits, based on manipulation of data as previously described. Most units permit operator adjustment of signal delay in small increments, measured in milliseconds, or thousandths of a second. Echo effects are characterized by a longer delay time as they simulate sound traveling and reflecting back across a greater distance. The length of delay and number of repeats can be readily varied on most reverberation devices.

Items of digital equipment capable of phase shifting, flanging, and chorusing effects are also marketed by a number of manufacturers. Because the sound produced by phase shifts primarily results from the comb filter phenomenon, many contemporary phasing and flanging units simply make use of comb filters to produce their effects. The filter's notch frequencies move across the spectrum at varying rates and with varying depths. Both the rates of movement and the depth are user adjustable. Examples of flanging and phase shifting can be heard on the accompanying CD-ROM.

When using digital workstations, processing via software keeps the production chain within the digital domain. In computer-based workstations, processing may be either fully integrated in the editing software or available in the form of plug-ins. When activated, either type of software processing places a virtual processor within the workstation's virtual production chain. Most DAW software such as ProTools, Adobe Audition, and similar production suites natively offer a set of processing options. These minimally include an assortment of reverberation, limiting, and equalization effects. Software processors can be set up to affect a single track, a multitrack recording, a group of tracks, or the stereo or monophonic output of a virtual mixer. The operator is normally given a broad range of processing choices by professional software packages.

Plug-in processors are installed in computers in the same way as any other software package. And like other software packages, plug-ins are produced for specific platforms, i.e., Microsoft Windows XP or Macintosh OS X. In addition, plug-ins must conform to one of several common formats that specify how the software integrates into workstations—VST, DirectX, TDM, and so on. Most DAW editors can work with only one or at most a few different specifications, so a plug-in intended for one software suite may or may not be usable with another package. Plug-in processors tend to offer more features than the built-in processing provided by DAW software, and they sometimes are specifically designed to recreate the effects produced by particular hardware-based processors; for instance, plate reverb or a particular model of limiter.

Once installed on a workstation, plug-ins usually are accessed directly through the DAW software. Typically, a drop-down menu provides engineers with a list of installed plug-in choices. By activating the selected plug-in, the producer can then set the range of features it provides. The flexibility afforded by software processors often exceeds their hardware counterparts. For example, reverberation controls may allow the operator to manipulate several different sound reflections, with each one afforded some degree of equalization and timing options.

Digital processing has given birth to a wide array of audio effects processors, many of which are variations on the ones just described. The **choruser** applies a short delay, in which the delay length is continuously modulated. The result is often described as a "shimmering" effect as changes in delay times introduce variable intermodulation components among signal frequencies. Intermodulation frequencies are induced when two input signals interact to generate new frequencies that are the sum and difference of input frequencies. For example, 440 Hz will interact with 660 Hz to create beat frequencies at 1,100 Hz and 220 Hz. Of course, real audio is likely to be highly complex, potentially containing many dozens of component frequencies interacting to form hundreds of other harmonics and overtones. Examples of chorusing are available on the accompanying CD-ROM.

Yet another type of digital device can alter signal input frequencies by effectively stretching or compressing the timeline along which an auditory event occurs. Among these devices are called the **Harmonizer** and the **pitch shifter**. In these, a digital circuit samples an incoming signal—for example, one at 440 Hz—and then raises or lowers the output frequency of the signal by a selected multiple so that the output frequency can be moved to 880 Hz, 1,340 Hz, and so on. Devices such as the Eventide Harmonizer claim to create natural-sounding pitch changes even on complex audio such as vocals. Some examples of pitch shifting can be heard on the accompanying CD-ROM.

The **multi-effects processor** remains popular today, especially in semiprofessional applications, and today many manufacturers offer examples. Some of these permit individual effects to be arranged and combined in user-defined virtual sequences. In other types of processors, these effects are arranged in a predetermined order. Still other multi-effects processors are exclusively intended to enhance vocals or "sweeten" the overall mix, and allow for very little user adjustment. (See Chapter 17, Editing, Sweetening, and Postproduction.)

NOISE REDUCTION PROCESSORS

With the advent of digital workstations and the consequent transition to recording and processing in virtual realms, new approaches to noise management in audio signals became possible. Whereas noise reduction was originally achieved using simple low-pass filter circuits, and then later was enhanced with various companding schemes, contemporary digital noise reduction methods based on mathematic models have become immensely powerful. Using advanced algorithms, they can eliminate only the exact unwanted signal components. Noise reduction plug-ins enable operators to sample a recorded segment, designate a portion of the signal as "noise," and then eradicate all instances of this sound from the signal. For example, to eliminate noise from a loud air-handler in a recording, an otherwise quiet passage in which only the air-handler sound is heard can be identified as noise. The processor then removes data from the waveform that exhibit those distinctive sounds. To achieve this result, software routines need only duplicate the noise signal digitally, shift the duplicated signal 180 degrees out of phase, and then recombine the two signals. This cancels the undesired signal and the offending noise vanishes.

FORENSIC SOUND

Effects algorithms such as ones providing noise reduction, filtering, and equalizing, as well as speed and pitch shifters, have greatly advanced a novel and potentially rewarding niche in the audio field known as "forensic sound." Forensic audio specialists use signal processing techniques to enhance sound quality in recordings employed in legal, criminal, and accident investigations. Recordings that are excessively noisy, such as aircraft cockpit or locomotive voice recordings can be significantly enhanced using noise reduction processing of the sort just described. Recordings of telephone or cellular telephone calls, audio from surveillance cameras, messages retrieved from poor-quality answering machines, or audio from clandestine recordings made during criminal investigations all may require intensive processing to be useful for law enforcement or accident investigations. The quality of enhanced recordings acceptable in these applications may not be high fidelity; the principal goal in forensic audio is usually just a high degree of intelligibility.

CHAPTER TEN

MONITORS AND LOUDSPEAKERS

The conversion of electrical energy into the physical movement of air molecules requires a transducer—most commonly, some type of **loudspeaker**. Unlike other types of audio technology, loudspeaker design has remained little changed for more than a century. Advances in materials have certainly been made, enclosure designs have been improved, and occasional esoteric design trends have emerged, but operational principles used by loudspeakers are basically the same as those first described in 1874 by Ernst Siemens.

Siemens filed his original patent for a moving **coil** transducer in 1874, but he did not add a parchment paper **diaphragm**, making his device capable of reproducing audio, until his 1877 patent. Siemens' design featured the exponentially flared "morning glory" trumpet, which had become a standard fixture on many mechanical speakers of the era. An example of an early flared trumpet horn is pictured in Figure 10.1.

A further design element, added by John Stroh in 1901, was a **paper cone** attached to the **voice coil** and terminating at the rim of the speaker in a pleat. The Stroh design includes most features of contemporary cone-type loudspeakers, while the Siemens design serves as the basis for most contemporary **compression** drivers employed on **horn loudspeakers**. The following discussion concerns these two important speaker designs.

DYNAMIC LOUDSPEAKERS

The majority of loudspeakers rely on the physics of electromagnetism. Because of their use of electromechanical principles, the Siemens and Stroh transducers are considered to be **dynamic loudspeakers**. When a magnet and a conductor wound into a coil are moved in close proximity to one another, a flow of electrons will be induced. This is the principle of the dynamic microphone, as discussed in Chapter 4, Microphones. If an alternating current passes through a coil loosely suspended within the field of a permanent magnet, the coil will be made to move. When the current phase polarizes the coil opposite to the polarity of the magnet

FIGURE 10.1   An early exponentially flared horn speaker.

the coil will be drawn to the magnet, but when the current produces the same polarity as the magnet the two will be repelled. In other words, when a coil and a magnet's north poles are aligned in the same direction, they will move apart. When their poles are aligned in the opposing direction, the two will be pulled toward each other. To observe how this simple concept is applied to loudspeakers, examine Figure 10.2. In this drawing the relationship between the voice coil, permanent magnet, and the diaphragm can be seen.

When Siemens added a diaphragm to his voice coil, he created a device capable of producing mechanical vibrations that could be transferred to adjacent air molecules. As the coil moved the diaphragm's membrane, it would set nearby air molecules into motion. To aid the process of moving more molecules, hence enhancing sound volume, Siemens added a trumpet-shaped horn. The exponential flare of the horn efficiently transferred energy to air by assisting in the propagation of sound waves. Technically speaking, the horn achieves this by better matching the **impedance** of the loudspeaker (diaphragm) with the air surrounding it.

The Siemens diaphragm was parchment, but present-day compression driver diaphragms are generally made of lightweight metals such as titanium. Figure 10.3 shows two modern diaphragm designs. Metal diaphragms have the durability to withstand physical stresses imposed by the high sound-pressure levels of contemporary sound reinforcement systems. Consider how powerful the vibrations must be across a one-inch diaphragm to generate the amplitudes required in long-throw sound systems used in large venues.

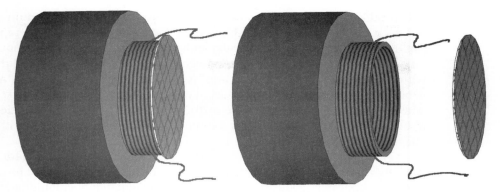

FIGURE 10.2 Modern loudspeakers are constructed of a permanent magnet into which a cylindrical voice coil is closely fitted. The voice coil is attached to a diaphragm or cone for production of sound.

While Siemens' ornate morning glory trumpet has been replaced by less decorative horn patterns, modern horn loudspeaker drivers are still designed along the same lines. The most popular approach incorporates an exponential flare rate from the horn's throat to its mouth. The flare rate describes the contour or cross-section expansion rate of the horn, moving from the diaphragm outward to the horn's mouth. Contemporary horns may otherwise incorporate flare rates described

FIGURE 10.3 Two modern metal horn diaphragms, capable of reproducing high sound-pressure levels at high frequencies when properly coupled to horns.

as conical, parabolic, tractix, and hyperbolic, but these are less common. The conical horn expands in a linear fashion; the exponential expands more rapidly near the horn mouth; the hyperbolic expands most rapidly nearest the mouth. Each type of flare offers a compromise between size, efficiency, and sound distortion, with conical being least efficient and hyperbolic being most efficient. As it happens, the exponential flare is the easiest to calculate, to fit into a compact package, and to minimize distortion while at the same time maximizing efficiency. The photograph in Figure 10.4 shows a contemporary exponential horn with a driver attached.

The paper cone John Stroh attached to the voice coil was a giant step forward in loudspeaker design. Today, speakers using paper cones are incorporated into just about every consumer audio device imaginable, including television receivers, computer speakers, bookshelf stereo speakers, automobile speakers, and portable radios. Paper cone speakers are used in a large majority of professional and commercial audio devices as well, including in-ceiling speakers, studio monitors, sound reinforcement speakers, and musical instrument amplifiers. By fixing a paper cone to the voice coil, a larger volume of air can be moved than if it were attached to the smaller diaphragm. The paper cone creates higher sound-pressure levels than a diaphragm, when considered independently of enclosures and horns, and provides wider frequency response than can be obtained from a diaphragm.

The basic components, then, of a modern cone speaker are a voice coil attached to a paper cone. The pleated outside edge of the cone is directly attached

FIGURE 10.4 Exponential horn with attached driver.

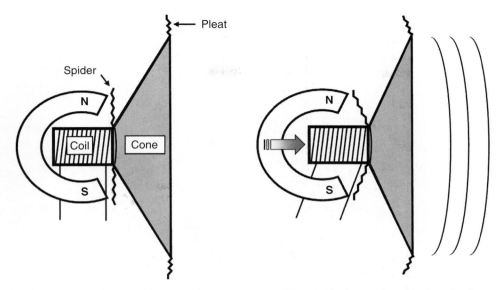

FIGURE 10.5   A cone driver speaker creates sound pressure waves when motions in the voice coil move the paper cone. Right: During a positive wave phase, the paper cone compresses the air molecules adjacent to it.

to a metal frame, known as a **basket**. The voice coil is also attached to the basket by another ring of pleated flexible material called the **spider**. Figure 10.5 presents the functional parts of a cone-type loudspeaker driver, and it shows how the cone flexes to create compression waves in air molecules near the cone.

In currently manufactured cone speakers, the **cone** might be made of paper, but it might also be metal, plastic, or other thin, lightweight materials. Cone speakers come in sizes ranging from approximately one inch to eighteen inches. The speaker cone shown in Figure 10.6 has been largely removed to reveal the basket; the spider, which has also been partially removed; and the voice coil.

Large speakers, those ten inches or more, are intended to reproduce low frequencies (below 200 Hz) and are called **woofers**. Small speakers, usually less than four inches, are used to reproduce high frequencies (above 3,000 Hz) and are called **tweeters**. Mid-range speakers reproduce frequencies between these ranges (from about 200 to about 3,000 Hz), and their sizes vary from four to ten inches in diameter. Very low frequencies (below 80 Hz) can be reproduced by **subwoofer** speakers greater than ten inches in diameter. Because of the limited frequency range of speakers, two or more of different sizes can be combined in a single enclosure to produce a wider frequency response.

In order to conserve space, and to reduce phase differences induced by multiple drivers, more than one voice coil and cone may be combined in a single frame. Examples include the **coaxial loudspeaker** and the triaxial loudspeaker. In a coaxial speaker, a tweeter or mid-range driver is placed in the center of a

FIGURE 10.6   A cone driver speaker with the cone and spider partially cut away to reveal the voice coil and permanent magnet from which the voice coil protudes.

woofer. Coaxial speakers are most common in automobile sound systems where space is at a premium, but they have also been incorporated into some studio monitors to provide a more phase-coherent single-point-source loudspeaker.

LOUDSPEAKER ENCLOSURES

In free space, a cone speaker is notoriously inefficient. It would require large amounts of power to create sufficiently high sound pressure levels for normal use. It is inefficient because sound waves coming from the rear of the cone are 180 degrees out of phase with sound waves from the front of the cone. In free space, these two waves tend to cancel each other out. All practical cone speaker systems place the driver in an enclosure that increases its efficiency by blocking the rearward sound waves from reaching the front of the cone. The surface onto which the cone speaker is mounted is called a **baffle**. The baffle could be a ceiling tile in a department store ceiling or a dashboard in an automobile, but baffles are most often thought of as the front surface of a conventional box-shaped speaker enclosure. Extending the baffle so that it completely separates the front and back of the speaker creates an enclosure called the **infinite baffle**. The ceiling speaker and the automobile dash-mounted speakers fit this description. Some acousticians argue that the infinite baffle is the most natural-sounding speaker design because it is uncolored by stray resonances in the box. But it is not efficient because it makes no use of the energy expended in the rearward sound waves.

A variant of the infinite baffle design folds the baffle back on itself to create a closed box. Such a design is usually referred to as an **acoustic suspension speaker** or enclosure. The acoustic suspension enclosure must be completely sealed so that no air can escape as the speaker operates. Air trapped inside the box acts as a spring against the reflex excursion of the speaker cone; the cone moves forward in its positive phase, rarefying the air inside the enclosure, but the cone does not rebound as far during the negative phase, because the air inside the box cushions it as it compresses. This damping increases slightly the efficiency of the speaker by reducing the distance of the cone's travel, but this is offset by the energy required to compress air trapped inside the box. As a result, the acoustic suspension-speaker system is less efficient than other designs. Even so, it is easy to construct and, with the proper transducer, can offer a reasonably flat frequency response. Figure 10.7 illustrates the acoustic suspension and several of the other most common speaker enclosure types described in this section.

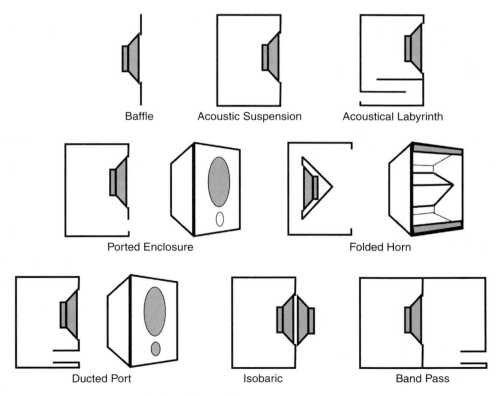

FIGURE 10.7   Some typical speaker enclosures.

A loudspeaker can be made to seem louder by boosting its bass response. The principle of equal loudness states that the human ear does not hear low frequencies as well as it does mid-range frequencies (see Chapter 1, Sound and Its Nature). Placing a loudspeaker in an enclosure that enhances bass response will, on its own, increase the speaker's apparent efficiency. One type of enclosure boosts low-frequency output by providing a pathway for sound waves from the cabinet interior to escape. The **bass reflex speaker** or **ported enclosure** places a hole in the baffle that allows some of the energy from the rear of the speaker to escape the box. As long as the low-frequency sound energy emanating from the port is in phase with the energy from the front of the speaker cone, the waves reinforce each other, raising the speaker's apparent loudness.

Bass reflex types include the tuned or ducted port, and the acoustic labyrinth or transmission line designs. The tuned port adds a length of duct, usually polyvinyl/chloride (PVC) pipe or heavy cardboard, to the port. The length and diameter of the duct effectively tunes the cabinet by determining which frequencies will be enhanced by waves from the rear of the driver. The acoustic labyrinth incorporates an extended path for energy from the rear of the cone driver to travel before exiting the cabinet. This is usually accomplished by placing a series of baffles in the cabinet that lengthen the path to the cabinet's port.

These types of enclosures represent the preponderance of ones found in professional use. Two variations do bear mentioning here, even though they are seldom seen. The **isobaric** enclosure includes the curious addition of a second loudspeaker attached "face-to-face" with an internal driver. The two speakers are connected out of phase, so that as the internal cone moves outward, away from its coil, the outer cone is moving toward its coil. The two cones work together in moving larger amounts of air. The isobaric enclosure has never gained much popularity because the external driver is exposed.

A **band pass** enclosure unites an acoustic suspension enclosure with a ported enclosure. The driver is mounted in the sealed enclosure, but pushes air into the ported enclosure, rather than into the open space of the room. Sound waves are reflected inside the ported enclosure and eventually exit through one or more ports. The effect of this is to augment the enclosure's bass response. Band pass designs can significantly extend the bass response of drivers of less than ten inches in compact speaker systems. Refer to Figure 10.7 for illustrations of these unusual enclosure types.

Because a horn speaker is the most efficient design in propagating sound waves, some loudspeaker designs combine the exponential dimensional relationship of a horn with a cone loudspeaker. Many subwoofers in sound reinforcement systems are actually horn-loaded enclosures. If the front of the speaker cone can be seen as the horn throat, and the front of the speaker cabinet as the horn mouth, and if the increase in area between those two openings reveals an exponential increase, the result constitutes an exponential flare horn.

The kinds of speaker drivers discussed here are ones most commonly seen today, and they represent the majority of speaker systems found in recording studios, radio, and live sound reinforcement. However, as noted at the outset, less-common types may be encountered. For example, there are alternatives to dynamic loudspeakers, just as there are to dynamic microphones.

ELECTROSTATIC LOUDSPEAKERS

An **electrostatic loudspeaker** is operationally similar to condenser microphones but, of course, much larger. Electrostatic speakers are constructed of a wide, thin, conductive diaphragm panel sandwiched between two electrically charged grids. The grids are connected in such a way that one is always electrically positive, and the other electrically negative. The conductive diaphragm is connected to an amplifier's output. As the phase of an audio signal applied to the diaphragm becomes positive, it will be attracted to the negative grid, and conversely, when the audio becomes negative, the diaphragm is attracted to the positive grid. The back-and-forth movements of the diaphragm set air molecules around it into motion. The diaphragm's low mass allows it to respond accurately to rapid changes in audio voltage, but because diaphragm movements are small, electrostatic speakers do not reproduce low frequencies well. Figure 10.8 shows the construction of an electrostatic speaker.

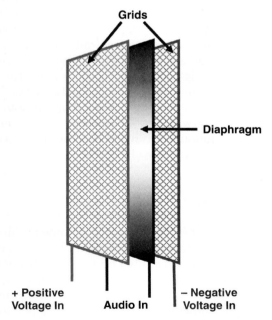

Grids

Diaphragm

+ Positive Voltage In   **Audio In**   **– Negative Voltage In**

FIGURE 10.8   The grids of the electrostatic speaker alternatively push and pull the diaphragm to create sound pressure waves.

PLANAR MAGNETIC LOUDSPEAKERS

Yet another unusual kind of driver parallels the design of ribbon microphones, and is variously known as a **planar magnetic** or **ribbon loudspeaker**. This type somewhat resembles the electrostatic speaker in its construction, employing a long, narrow, thin, metallic ribbon suspended between two permanent magnetic poles rather than electrostatic grids. As the audio signal applied to the ribbon alternates between positive and negative phases, the ribbon is successively attracted to and repelled by the fields of the magnets. Ribbon speakers are incapable of moving large volumes of air, so their use is limited to applications as high-frequency transducers or tweeters.

ACTIVE LOUDSPEAKERS

Servo speakers are among the most complex speaker systems ever devised. Employing feedback loops, these types can sometimes be found in **active** or powered **loudspeaker systems**, in which an amplifier is made an integral part of the speaker system. Servo circuits are intended to reduce distortion, especially in low-frequency woofers and subwoofers where waveform errors can be high due to the use of large magnets and heavy cones. These are needed to generate the long excursions required for low frequencies at high sound-pressure levels. However, the weight of their cones leads to lower compliance; that is, they do not respond as quickly.

In active loudspeaker systems, a line level signal from a mixing console or other audio source is sent directly to an amplifier within a speaker enclosure. Servo systems contain a device called an accelerometer that measures cone movement. The output of the accelerometer is sent to a comparator chip that compares the travel of the cone with the movement needed to accurately represent the amplifier's input signal. The amplifier's output is then lowered or raised to compensate for errors found in the cone's movement. Of course, this happens within microseconds and is performed continuously. As might be expected, servo speaker systems are expensive and for this reason not widely used.

CROSSOVER NETWORKS

As mentioned previously, most practical speaker systems combine multiple drivers within a single enclosure to more accurately reproduce the whole audio frequency range. The short, rapid movements required at high frequencies cannot be reproduced by a heavy, fifteen-inch woofer, and a two-inch tweeter cannot achieve the large excursions required to propagate low frequencies. Speaker systems are described as **two-way**, *three-way,* and sometimes *four-* or *five-way,* depending on the number of drivers contained in each enclosure. The audio signal sent to the speaker system is split into selected bands of frequencies by a **crossover network**

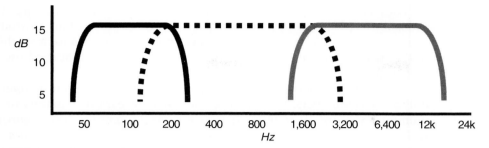

FIGURE 10.9   A three-way crossover network divides audio signals into three bands that can be routed to three separate amplifiers.

before routing to appropriate drivers. Figure 10.9 illustrates how a three-way crossover divides the audio spectrum, with frequencies below 200 Hz routed to the subwoofer, 200 Hz to 3,000 Hz routed to the mid-range, and frequencies above 3,000 Hz routed to the tweeter.

Crossovers can be either **passive** or active types. Passive crossovers that require no external power are based around inductor and capacitor networks. As frequencies rise, inductors tend to oppose the passage of energy while capacitors' opposition tends to decline. In combination, the values of network components will determine their cutoff frequencies. Active crossovers are tunable filters that can be adjusted by users to select cutoff crossover frequencies or points. Suitable crossover points depend on the response characteristics of the speakers employed. For instance, eighteen-inch subwoofers produced by different manufacturers may have specifications that are quite dissimilar, and thus require different cutoff frequencies. Speaker systems ordinarily place crossover networks within the speaker enclosure, but active crossovers can be outboard devices.

Live sound reinforcement systems and certain high-quality studio monitor systems also use **bi-amplification** (bi-amping) or **tri-amplification** (tri-amping). In a bi- or tri-amplified system, the outputs of the active crossover network are routed to separate amplifiers that are connected to individual speakers. To obtain a reasonably flat overall response, high frequencies may require more amplification than the mid-range amplifier—as proposed by the principle of equal loudness—and the low frequency amplifier may require substantially more power than the mid-range amplifier. Chapter 12, Live Sound Reinforcement, describes these systems more fully.

LOUDSPEAKER MONITOR PLACEMENT

Placement of loudspeakers within a room or venue, and in relation to the listener, is vitally important to their perceived sound characteristics. For stereo playback, the usual arrangement has been to position the two speakers and the listener at the corners of an isosceles or equilateral triangle, as illustrated in Figure 10.10. The availability of space within rooms where speakers are located may introduce constraints on alignment of the triangle. Nevertheless, if the distance from the listener to the two monitors is not the same, it should be obvious that the closer speaker will always appear to be louder.

The distance between a loudspeaker and the listener can be many feet or only a few centimeters. They can even be inside the ear canal, in the case of in-ear monitors. Monitor placement schemes in control rooms can be categorized as **far field**, *midfield*, **near field**, and even *ultra-near field*. *Far, mid,* and *near* in these terms refer to the distance between the monitor and the control room operator. Placing the monitor near the listener ensures that the sounds arriving at the operator's ears will come predominantly directly from the speaker and not by reflection from surrounding walls. This typically means a distance of three to six feet. For decades, recording studio control rooms commonly made use of far-field monitors. Far-field monitors impose several limitations and design considerations. Far-field monitors are larger; to power them, greater amplification is required. Naturally, the sound of the far-field monitor is greatly affected by character of the room. As the length of the path traveled by sound from the speaker to the listener increases, opportunities for coloration and standing waves grow greater.

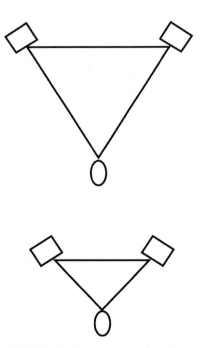

FIGURE 10.10   A triangular placement of listener and stereo speakers helps ensure accurate stereo imaging.

In recent years, near-field monitors have become favored in many control rooms. Chief among reasons for their popularity is the smaller size of enclosures, something that matters a great deal in project and home studios. Also, near-field monitors minimize the effect of the room on sound. Again, this is particularly important in project and home studios, which tend to have minimal acoustic treatment. With near-field monitoring, room dimensions and geometry and surface treatments are much less critical. The introduction of multiple-point-source speaker systems (surround sound) has also reshaped thinking about monitoring and mixing.

Some recording engineers and acousticians contend that as distances between the speaker and the listener are reduced there is a point of diminishing return. They believe that it is possible to be too close to a monitor and that locating oneself too close to a monitor creates an unnatural listening environment. For similar reasons, no one recommends mixing with headphones—unless the audience will only listen with headphones. Binaural recordings, made using a dummy head mic system and intended to be heard on headphones, are one example.

LIVE SOUND SPEAKER PLACEMENT

Speaker placements in live sound reinforcement have also undergone change over the years. In the early decades of the field's development, conventional wisdom called for two clusters or stacks of speakers on opposite sides of the stage. Perhaps this was an outgrowth of the rising popularity of stereo recording, the growth of which paralleled that of live sound reinforcement. Use of this layout did not necessarily mean that signals were in stereo, only that the speaker placement emulated stereo placement. During the late 1970s and early 1980s, some touring companies even experimented with **quadraphonic** sound systems, in which left-rear and right-rear clusters complemented traditional left-front and right-front speaker clusters.

Within the past two decades, many touring sound companies, sound contractors, and artists have reconfigured their systems into **single-point arrays**. These consist of a single cluster of speakers usually "flown" above center stage. Proponents of this arrangement maintain that it reduces phase coloration problems associated with signals arriving at the listening position from two different sources in differing phase relationships. Moreover, they suggest, this configuration offers more evenly distributed coverage in most venues, with fewer dead spots. Many live shows are "mixed in mono," even though speakers may be in left-right clusters, so switching to a single-point array presents no serious problems for the front-of-house engineer. However, the space and rigging requirements for flown systems make this approach impractical for all but the largest shows and systems. Figure 10.11 illustrates the single-point array from both a side and front view. Figure 10.12 shows the more traditional side cluster configuration for arranging speakers for live sound.

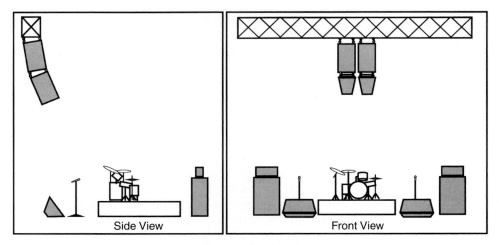

FIGURE 10.11   Two views of a single-point loudspeaker array.

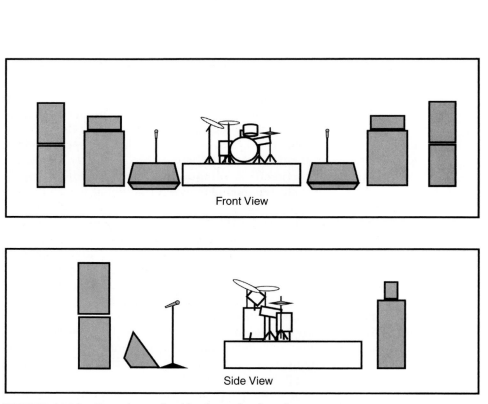

FIGURE 10.12   Two views of a side-cluster loudspeaker array.

LISTENER FATIGUE

Audiologists suggest that single-point arrays may also reduce **listener fatigue**. Listener fatigue is a condition in which listeners experience symptoms of physical exhaustion, triggered by sound abnormalities. The symptoms stem from high sound levels, distorted audio, or inconsistencies between the apparent sound sources and the actual sound sources. According to theory, the more difficulty the brain has to match a sound to its source, the more quickly the listener tires. Listener fatigue may also be set off by a particular speaker type or enclosure design. For instance, some listeners report that speaker enclosures having weak low-frequency response induce fatigue, and many people report fatigue after extended periods of headphone use.

One thing is certain: sustained exposure to moderately high sound pressure levels of above 80 dB will rapidly degrade a listener's hearing. Those high and low frequencies to which the human ear is least sensitive are the first to be affected by excessive sound levels. Exposure to program material at over 80 dB for thirty minutes or more will further diminish the ear's sensitivity to high and low frequencies. As a result of this desensitization, audio operators are likely to respond incorrectly by boosting equalization at upper and lower frequencies. A common practice among recording engineers is to take a listening break at least on an hourly basis during recording and mixing sessions to forestall the effects of listening fatigue.

SOME TECHNICAL CONSIDERATIONS

Technical specifications for loudspeakers rate their characteristic impedance or loads, and power handling capabilities. These are primarily derived from a speaker's design and construction. Generally, at the same input level, a loudspeaker having lower impedance will produce a higher sound pressure level than a higher-impedance loudspeaker. Total impedance of the loudspeaker is determined by three factors: The first factor is the electrical **impedance** of a loudspeaker's voice-coil, as determined by its combined resistance, inductance, and capacitance. Electrical impedance can be thought of as a speaker's opposition to alternating current flow. The second factor is the driver's *mechanical impedance,* largely determined by the cone's opposition to movement. This is affected by its mass, damping, and stiffness. Finally, the third factor contributing to speaker impedance is the speaker's *acoustic radiation impedance,* which is how efficiently the driver moves air, and this is determined by the physical dimensions and structure of the loudspeaker system.

Even though loudspeakers are rated by manufacturers at particular characteristic impedances, usually 4, 8, or 16 ohms, this figure is not constant across the range of audio frequencies. A speaker's impedance is more properly referred to as its *nominal* impedance, meaning a single value that summarizes its broadly fluctuating characteristics. Variations in a speaker's actual impedance at any frequency arise from myriad factors such as the driver, the enclosure, and speaker placement. The match between a speaker's impedance and the power amplifier's output impedance is important. Mismatches will result in compromised performance of the sound system and place stress on the amplifier. At the same power level, an amplifier that has a low impedance output is capable of delivering greater current at a lower voltage than one having high output impedance. A higher-impedance amplifier provides higher voltage but at a lower current.

The most efficient transfer of power to the speaker system occurs when the electrical impedance of the speaker presents a load equal to the amplifier's output impedance. This ensures that the current and voltages delivered to the driver fall within the design parameters. As the impedance of a loudspeaker declines, an amplifier must provide greater output current. In extreme cases, amplifier failure can occur when it is unable to supply the current demanded. A loudspeaker system's combined load of 2 ohms places most amplifiers at their design limits. Fortunately there are wiring techniques and related measures that can be taken to manage impedance loads (see Chapter 12, Live Sound Reinforcement).

Matching loudspeaker power ratings to the proper amplifier power output is more significant than it might seem. If amplifier power output is much higher than speakers are rated to accept, speaker damage can easily occur. When the amplifier delivers too much power to the speaker, the voice coil is likely to fail. The light gauge wires used in coil windings can be melted due to excessive current flow. In multidriver systems, the tweeter is most prone to damage in this way because it has the smallest voice coil with the thinnest wires. Another danger of too much amplifier power is damage resulting from extreme cone excursion. This occurs when the cone travels beyond its design limits and dislodges the voice coil from the speaker magnet or tears the cone.

Less obvious problems can arise if an amplifier lacks sufficient power to generate desired sound-pressure levels. It would seem that connecting a 50 W amplifier to a 200 W speaker would not pose a risk to the speaker. However, if the operator attempts to operate the amplifier beyond its maximum designed ratings in order to obtain high sound-pressure levels, problems will definitely result. Overdriving the amplifier will cause it to clip the signal, and this raises its high-frequency content dramatically. Distortions created by clipping will stress high-frequency drivers and promote their failure.

CHAPTER ELEVEN

TRANSPORTING SOUNDS

This chapter is devoted to technologies used to move sounds from one location to another. The following discussion focuses not only on technologies and equipment used to transport audio signals over long distances but also those available to send signals short distances within a production setting.

COPPER WIRE TECHNOLOGY

The earliest methods used to transport sounds over long distances were by-products of telephony's development. Telephone lines in the late 1800s carried analog audio signals on a single pair of copper wires—one pair for each signal or circuit. Those wires were unshielded, bare copper wires strung from poles. That technology had limitations. Signal attenuation when traversing long distances could be offset somewhat by increasing the wire diameter and by the use of loading coils. In the mid-1910s, vacuum-tube repeater amplifiers were added to telephone networks, making it possible for signals to travel much longer distances. Yet these lines offered bandwidths of only about 3,500 Hz, just enough for essential frequencies of the human voice. Other imaginative engineering techniques, such as circuits that superimposed audio signals on carrier currents, tripled the bandwidth capacity of transmission lines.

Greater bandwidth was important not only to meet demand for higher telephone-call capacity, but also to make the system suitable for the radio networks that were just being formed. These radio broadcasting networks adopted telephone circuits to connect themselves with individual stations. Early broadcasters used amplitude modulation transmitters so they needed an audio frequency bandwidth of merely 5 kHz, a requirement easily accommodated by telephone circuit technology of the 1920s and 1930s.

Copper wire technology continues to be the backbone of much the telephone companies' (**telcos**) network infrastructure today. Copper wires—commonly called **twisted pair** lines—carry voice and data signals within homes and buildings, and then move them onward to the telco central office where twisted pair circuits interface with other types of transmission lines. Twisted pair lines remain in use because they are inexpensive and easy to install. A multitude of technological advances have been developed to support the use of twisted pair lines, including ones to carry wide-band audio signals, broadband data, and even video signals.

Beginning in the late 1930s, increasing demand for telephone circuits and the emergence of embryonic television broadcasting contributed to the parallel development of **coaxial cable** and **microwave** transmission technologies. The bandwidth of coaxial cable transmission lines was far greater than that of twisted pairs. Coaxial cable derives its name from the fact that the two conductors have the same axis; the center conductor is made of an insulated stranded or solid copper wire around which a web of very thin braided wires is wrapped. The first coaxial-cable transmission lines were able to carry a single television channel occupying a bandwidth of 6 MHz—as much as three hundred times greater than audio bandwidth. By the 1970s, technical advances enabled a coaxial transmission line to carry more than two hundred television signals. These lines are easily capable of delivering FM-quality stereo audio signals containing frequencies up to 15 kHz over long distances.

MICROWAVE TECHNOLOGY

The development of microwave technology was driven both by the need to accommodate television signals and by the need for more and more telephone circuits. Early microwave circuits came into operation by the late 1940s when TV networks began relaying programs across the United States. Microwave transmission derives its name from the extremely short wavelengths of the radio frequency (RF) **carrier** onto which audio signals are modulated (30 cm to 1 mm).

Microwave frequencies are in the range of 1,000 MHz to 50 GHz, and occupy the super-high frequency (SHF) and extremely high frequency (EHF) bands of the RF spectrum. Microwave transmissions make use of radio waves at such high frequencies that they travel like light waves—in a straight line. Relay stations generate powerful microwave carrier waves on which a large amount of multiplexed data can be modulated. This can amount to thousands of telephone circuits, hundreds of wide-band audio circuits, or several television signals. Microwave signals are transmitted and received using narrowly focused parabolic dish antennas similar to those used for satellite reception. Given adequate power and an unobstructed path, microwaves are capable of traveling distances up to about forty miles.

Another of the incentives behind the growth of microwave transmission networks was the need to transmit telephone signals over rough terrain or to remote areas where stringing copper lines would be difficult or expensive. Microwave transmission is useful in crossing bodies of water and in reducing construction costs of communication circuits in congested urban environments. Even in flat open terrain, costs of installing a few microwave relay stations can be far less than the expense of installing hundreds of miles of cable.

SATELLITE TECHNOLOGY

The step forward in the transport of audio signals was literally a giant leap—into outer space. In 1945, Arthur C. Clarke published a technical article in which he suggested that earth-orbiting artificial satellites could receive microwave signals beamed from a directional antenna on the planet's surface, and these signals could then be transmitted back to the earth using the satellite's onboard directional antenna. A receiving antenna located anywhere within the satellite's **footprint** (that portion of the earth's surface covered by a satellite signal) could pick up the relay transmissions. Clarke's article outlined a plan for placing satellites in an orbit that causes them to circle the earth at the same speed as the earth's rotation, so that to an observer on earth the satellite would appear to be stationary in the sky. Satellites like this are termed **geosynchronous**. Because satellites appear to be suspended in a fixed spot above the earth, they are labeled **geostationary**. In 1965, twenty years after Clarke's proposal was published, the United States' first domestic communication satellites were launched. Presently, two types of communication satellite systems are in use—the **DBS** (direct broadcast satellite) system and the fixed service system (FSS). FSS satellites operate in the microwave C-Band portion of the electromagnetic spectrum, from 4 to 6 GHz, and also in the lower portion of the Ku-Band, from 12 to 18 GHz. C-Band satellites operate with comparatively low power, and therefore require larger receiving antennas known as television receive only (**TVRO**) dishes. A C-Band TVRO dish is typically twelve feet in diameter or larger. C-Band transmissions are now used only for distribution of television feeds, teleconferencing, telephone circuits, and other private/leased services. Full transmission systems can connect satellite relays with terrestrial microwave networks as illustrated by the drawing in Figure 11.1.

DBS satellites operate in the upper part of the Ku-Band, with much higher power than C-Band satellites. The higher-powered transponder on board satellites allows receiving antennas on earth to be much smaller than FSS dishes. A DBS dish can be as small as eighteen to twenty-four inches in diameter. Ku-Band satellites also relay their signals using a digital compression scheme that allows more signals to be accommodated in their allocated bandwidth. DBS transmissions are generally intended for public consumption through subscription services such as DirecTV or Dish Network. DBS satellites are also used to deliver music channels for both private and commercial use. Companies like Muzak and AEI deliver licensed background and foreground digital music programming to commercial clients using Ku-Band satellites. Recently, some providers have begun to deliver broadband Internet services to residential customers via satellite, and a number of other types of digital data streams are transmitted via DBS as well. For example, the Associated Press delivers text, audio, and video news feeds to member news outlets via satellite.

One popular application for satellite communication technology is **satellite radio**. There are currently three companies engaged in space-based digital audio radio service: Sirius Satellite Radio, XM Satellite Radio, and World Space. Each

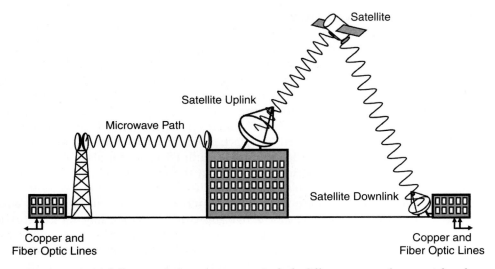

FIGURE 11.1 A full transmission system may include different types of terrestrial and satellite relays.

provider maintains satellite services in the S-Band and L-Band segments of the RF spectrum to deliver a multiplexed digital signal to subscribers. XM uses geostationary satellites, but Sirius does not, relying instead on three satellites in elliptical orbits so that at least two are usable in North America at any moment. World Space uses geostationary satellites to cover Europe/Africa, Asia, and South America. Satellite radio also relies on ground-based repeaters to provide uninterrupted coverage in selected urban areas.

FIBER OPTIC TECHNOLOGY

In the 1980s, another technology for voice and data traffic across telephone networks arose—**optical fiber**. Optical transmission of signals through strands of fiber relies on a source, either laser or LED, to relay information digitally coded as light pulses. Optical fiber is formed from highly purified optical glass drawn into a single continuous strand the diameter of a human hair. To contain light as it travels along extended lengths of fiber, a sheath of reflective cladding is wrapped around the optical fiber's exterior. This technique, known as total internal reflection, significantly reduces light attenuation in traveling through the fiber. Total internal reflection also allows light to be carried efficiently even if the fiber is gently bent or curved, as shown in Figure 11.2. Just as with wired circuits, an optical signal incurs losses as it travels, and if its pathway is lengthy, repeaters will be needed to boost its intensity. The signal capacity of fiber optic systems can be enormous, greater than any

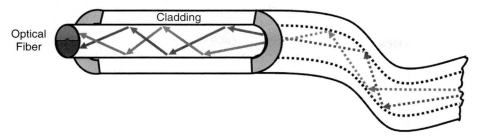

FIGURE 11.2   Optical fiber that uses cladding will efficiently conduct light waves around bends or curves in the cable by reflecting light inward.

copper wire technology. By making use of light of differing wavelengths, many separate signals can be carried within a single fiber strand. Optical cables are now commonly used for interconnecting pieces of digital audio equipment.

DIGITAL WIRED TECHNOLOGY

Any of the technologies described in the preceding pages—twisted pair, coaxial cable, microwave relay, satellite, and optical fiber—might be included in any single voice/data telephone circuit completed across the nation or around the globe. Starting in the 1960s, telcos began converting their central offices to the packet-based digital systems that now handle all telephone call routing and switching. Still today, however, from the central office to the customer, telcos mainly depends on analog twisted-pair connections, colloquially known as **POTS** ("plain old telephone service").

To enlarge the bandwidth of POTS technology in the 1980, telcos devised a new protocol and new systems of digital switching equipment. That system, called **ISDN**, or Integrated Services Digital Network, digitizes the analog signal at the point where a customer connects with the telco network. An ISDN terminal acts somewhat like a modem, converting analog audio signals to the ISDN digital packet protocol. ISDN circuits are capable of transmitting digitized versions of voice, wide band audio, text, graphics, or even video data simultaneously over standard twisted pair—wiring telephone networks. ISDN transfers should not be confused with Internet transfer; ISDN signals are exchanged between ISDN terminals, not computers. ISDN lines can be aggregated to increase bandwidth even further. ISDN circuits are available to telco customers on a monthly lease basis, and although they are normally too expensive for consumer use, many recording studios and broadcast stations make use of them. The bandwidth and transfer rates afforded by ISDN circuits allow studios to transmit **CD-quality** digital signals in real time.

Recording studios often employ ISDN hookups to bring distant performers and talent into productions. This is particularly common among recording studios that specialize in radio or television commercial productions. For instance, via ISDN links a vocal performer physically present in a New York City recording studio can overdub vocal tracks in a session at a post-production house in Los Angeles.

INTERNET-BASED TECHNOLOGY

If files are exchanged over the Internet, data packets actually route between computers across some of the same telephone networks just described. Generally speaking, audio can be transported across the Internet in one of two ways—it can be sent as a complete file, just as in transferring any other sort of computer file, or it can be sent as a stream with data encoded and transferred continuously in near-real time. Note that streaming media content sent over the Web is not truly in real time, as it is with ISDN; data instead is buffered, as described elsewhere in this chapter.

When files or data streams travel between computers across the Web they must share a common set of rules for their exchanges. These rules are called protocols. Transactions on the Internet employ twin all-purpose protocols—transmission control protocol and Internet protocol—together they are referred to as **TCP/IP**. For any two computers to communicate requires that they use TCP/IP. Additional protocols are needed to exchange certain types of data. Simple mail transfer protocol (**SMTP**) and post office protocol (**POP**) handle transfers of email, and hypertext transfer protocol (http) is used for transferring Web page files.

Large data files can be exchanged using the file transfer protocol (**FTP**), which is based on the TCP/IP protocol. FTP file transfers use **file servers** as intermediary computers on which files are stored for relay to other computers. For security purposes, FTP servers may be set up with password protection for file access. FTP software affords users the ability to copy, delete, rename, and move files while they reside on the server. In a typical FTP exchange, a user may **upload** files to a secure FTP server using appropriate software. Other users can **download** those files from the server at any time afterward, as shown in Figure 11.3. Many studios and production houses use FTP sites to make audio files available to clients. A radio commercial recorded in Los Angeles can be posted to an FTP server in Seattle, from which an ad agency in Detroit can download the file for duplication and distribution to radio stations by means of yet another server.

In transporting audio across the Internet, it is often preferable to send compressed files in order to shorten transfer times. An audio **codec** (i.e., coder-decoder) is a piece of software that transparently compresses and decompresses audio files according to one of several file or streaming audio formats. To prepare a file for Internet transfer, care must be used in selecting a codec or compression scheme, as the choice will determine file size, transfer rate, file type, and data rate. For professional audio files, and especially for music, where sound quality is far more important than transfer speed, no compression should be applied. This kind of file is sometimes referred to as using a 1:1 compression ratio or a lossless com-

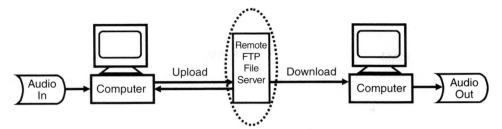

FIGURE 11.3   File exchanges employing FTP across a network are possible through the use of a file server.

pression. Even though the amount of time needed to upload/download files using uncompressed formats will be greater, they are favored among pro-audio users because of their superior frequency response and dynamic range. Two widely used audio codecs that support 1:1 compression are **AIFF** and **WAV**, and are identified with file extensions .aiff and .wav. Both are pulse code modulation (**PCM**) formats, and both support a range of bit rates, most often 16-bit. The AIFF compression format was developed for use on the Macintosh and Unix platforms, while Microsoft developed the WAV format. Both file types can be accommodated in most professional sound editing software. The AU (.au) format developed by Sun Micro Systems also provides uncompressed audio but is not used so extensively.

As an alternative to uncompressed files there is the **lossless codec**. Examples include Monkey Audio, FLAC, Wavepack, and Windows Media Audio, among others. Although lossless, these codecs are still compressed. These formats use 2:1 compression, but provide audio very close to the original quality after files are decompressed. These codecs are used occasionally for transporting and storing professional audio music files, in cases where bandwidth and transfer time are major considerations. The lossier formats use compression ratios from 4:1 (e.g., MPEG1) to as high as 53:1 (e.g., VOX). These codecs are rarely encountered in professional audio work, but are popular among consumers who desire faster transfer rates or higher storage capacities. Some of these formats are also utilized in specialized applications such as digital and Internet telephony and imbedded Web page audio. The speed and size gains of these highly compressed codecs are realized at a cost of reduced fidelity of course.

Growing in importance as a way of moving content over the Internet is *streaming audio.* As the term implies, streaming audio is sent continuously and played immediately as it is received, somewhat like a broadcast signal. There is at least one significant conceptual difference between streamed Web content and broadcast signals; however, listeners may interactively choose specific Web program material they wish to hear. To produce streaming audio, a digitized audio signal must first be compressed before sending it to listeners in a steady flow of small packets. On reception, the packets are decompressed and converted back into analog audio. A problem in sending audio over the Network is that data flow is never perfectly even. Because of traffic congestion, delays in the arrival of occasional packets are inevitable. To accommodate this problem, software used to process the incoming packets must employ a **buffer** to store data for a short time before playback. Buffering causes the stop-and-start flow of data to be transformed into a regular, evenly paced stream.

Streaming content relies on special sets of protocols that handle data transfer differently than ones used to send complete files. Streaming protocols such as user datagram (UD) and RealTime streaming protocol (RTP) send data continuously, rather than interrupting a transfer when data integrity is lost. Protocols such as http or TCP require resending data that is lost or corrupted. When a streaming protocol is combined with the least **lossy codec** available, reasonably good-quality audio can be delivered. Presently, there are more than a dozen different formats for streaming audio over the Web, with some of the most popular being proprietary solutions like RealAudio, Windows Media, and Apple's Quick Time. These streaming applications require that listeners use a matching player to hear the streaming content. MP3 and other nonproprietary formats are popular among file-sharing enthusiasts, as this codec offers acceptable audio that can transfer easily across a variety of platforms. Streaming solutions like Clipstream do not require specialized software, but play the content through a Web browser using Java. Still other applications, including Liquid Audio, offer a secure environment solution designed to sell audio content across the Web.

PODCASTING

A way of transporting sound across the Internet that has gained great popularity is **podcasting**, a method that takes its name from the Apple iPod. When both over-the-air broadcasters and independent content creators began encoding radio shows as MP3 files for distribution on the Web, it became known as podcasting, in recognition of the iPod's market dominance. The iPod is a trade name applied to several models of small, portable MP3 audio players manufactured by Apple Computer. Other companies manufacture portable MP3 audio playback units sold under their own trade names. The speed and scale of podcasting's acceptance as a delivery system for audio is nearly unrivaled in home entertainment history. Within a very short time, podcasting has emerged as a worldwide popular-culture phenomenon.

Podcasting begins as an audio recording of either an on-air broadcast program or one intended for direct distribution in this format. Although most broadcast content comes from radio, a few television stations deliver audio portions of their televised newscasts as podcasts. The producer of podcast programs must first convert the original audio in whatever format it is stored to an MP3 format, setting compression levels for suitable quality and file size. Next, the podcaster uploads the MP3 file to an Internet website from which listeners are able to download the program for playback on their MP3 players. Those who do not own

MP3 players can employ freely available software to hear programs played on their computers. Alternatively, the podcaster may email the MP3 files to a list of subscribers. Podcast enthusiasts may install software, called podcatchers, on their Internet-connected computers. Podcatchers automatically download selected podcasts as they become available. A few podcasts require a small subscription fee, but most are available at no charge.

Of course, podcasts are subject to the same copyright and music licensing rules as any other distribution or public-performance technology. Many would-be podcasters have been stymied by legal requirements for the use of copyrighted and licensed music. Broadcast stations pay substantial licensing fees to the American Society of Composers, Authors, and Publishers (**ASCAP**) and/or Broadcast Music Incorporated (**BMI**) to broadcast any material in their libraries, but most independent podcasters find those fees too costly. The alternatives are to include only public domain content, to include content created by the podcaster, or to use musical passages that are less than forty seconds in length so that excerpts fall within fair-use provisions of copyright laws.

Despite such constraints, proponents often argue that podcasting holds the promise of democratizing audio entertainment by providing an easy means for any producer to share their recordings with the world. On the other hand, critics have argued that the ease of creation and distribution has led to reduced quality standards so that listeners must apply their own filters to screen out poorly produced, inaccurate, or other kinds of substandard programming.

In summary, to prepare audio material for transport across the Internet a number of considerations beyond choosing appropriate software must be weighed. Among these are the desired audio quality and compression level, bandwidth requirements, cost, browser compatibility, and the need for interactivity.

BROADCAST TECHNOLOGY

Perhaps the most familiar way of sending audio over long distances is by radio **broadcasting**. On AM bands, radio signals can be received over paths of up to two hundred miles in the daytime, but under favorable nighttime conditions, these broadcasts can be received over distances of thousands of miles. FM broadcasts are usually limited to distances of one hundred miles or less.

When an analog audio waveform is impressed on a radio signal, the process is called **modulation**. Modulation is a way of placing an audio signal on a radio signal so that it will be carried along as the radio wave moves away from the transmitter. The radio wave is called a carrier because it is responsible for conveying the modulated signal. There are a number of ways to place information on radio waves, the most common being **amplitude modulation** (AM) and frequency modulation (FM).

In the early decades of broadcasting, amplitude modulation was the only means of transmitting sound by radio. In AM, an audio signal's waveform controls the RF signal's level or amplitude. At any instant in time the strength of an AM carrier depends on the audio voltage being sent to the transmitter. When the audio waveform is at a positive peak, the radio signal output will also be at a peak. Similarly, when the audio wave is at a negative point, the radio signal will be at a minimum. Figure 11.4 illustrates the waveform one might see of a 800 kHz signal amplitude modulated by a 1,000 Hz audio signal.

Audio frequency response in AM broadcasting is significantly more restricted than in FM broadcasting, but the limitations are a function of channel bandwidth and not of the modulation process. In the United States, AM radio broadcast frequencies are assigned from 530 kHz to 1,705 kHz. AM stations are provided a channel of 10 kHz, made up of 5 kHz on each side of their assigned frequencies. For a 10 kHz bandwidth, the maximum modulating audio frequency can be only 5 kHz. AM transmitters usually roll off high frequencies around 5 kHz to 7.5 kHz.

Impressing audio on a radio signal by shifting its carrier frequency is called frequency modulation (**FM**). This technique was demonstrated by Edwin Armstrong in the 1930s but did not gain wide acceptance until more than two decades later. In FM broadcasting, an analog audio waveform controls the carrier's frequency at any instant. In FM broadcasts, the higher the peak of an audio waveform, the greater the carrier shifts away from a station's assigned carrier

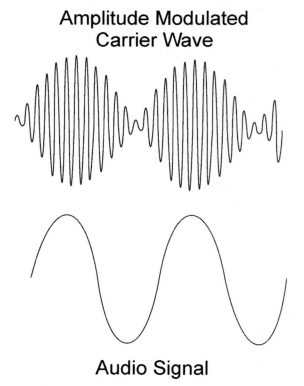

Amplitude Modulated Carrier Wave

Audio Signal

FIGURE 11.4   Amplitude modulation.

frequency. A carrier frequency's shift is thus determined by the audio waveform's voltage at any moment. Figure 11.5 shows how the frequency of the carrier follows the changing voltage of a modulating waveform. The frequency variation is exaggerated in this drawing to make the effect more visible. Observe that the FM carrier waveform height never changes, and thus has no modulation of its amplitude. Stereo FM broadcasting, in which separate left and right signals are contained on a single RF carrier, is achieved through multiplexing. Multiplexing is the technique of distributing two or more signals through a single channel. FM stereo transmitters place left and right channel signals on the carrier separately from the monophonic signals. This allows nonstereo receivers to pick up stereo transmissions as monophonic broadcasts.

FM broadcast stations are assigned transmitting frequencies between 88 and 108 MHz. Each FM station receives an authorization to broadcast in a channel 200 kHz wide. In FM, the frequency bandwidth of the audio signal can extend to 15,000 Hz. Above 15,000 Hz, frequencies are cut off sharply by the transmitter.

It is possible to transmit audio signals by radio in digital form. Digital transmission has all the advantages of digital recordings: low distortion, excellent signal-to-noise ratios, and the absence of degradation over a transmission path. In theory, audio quality can be comparable to CDs. Broadcasts in digital formats were authorized by the **FCC** in 2002, and iBiquity Digital Corporation has been licensing its technology to U.S. radio stations since 2003. It claims hundreds of U.S. stations are now broadcasting digitally.

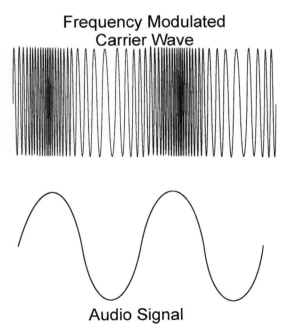

FIGURE 11.5 Frequency modulation.

How a signal is to be transported is important to the producer in several ways. First, the producer needs to be aware of the bandwidth of the audio signal that will be permitted by the method employed. For instance, audio material for AM broadcasting must have a natural and acceptable sound when limited to a bandwidth of 5 kHz, since frequencies higher than this will not be heard by listeners. Because FM stations allow a 15 kHz bandwidth, this restriction is not so important. FM transmissions are based on on variations of carrier frequency and so interference from amplitude variations caused by electrical static has very little effect. AM, on the other hand, is definitely prone to this problem. This is most obvious when listening to an AM station during a lightning storm; the static crashes can rise to a level sufficient to drown out the station. An additional consideration has to do with the processing that most stations apply when broadcasting their programs. Stations commonly use limiters to prevent transmitters from being overmodulated. However, the effect of this is to compress the dynamics of the audio. As a result, audio material intended for broadcast needs to be produced in such a way that the compressed sound still seems natural and correct.

CHAPTER TWELVE

LIVE SOUND REINFORCEMENT

The term live sound reinforcement is used to describe the practice of boosting sounds in performances that are not loud enough to be heard well by live audiences. Engineers engaged in this line of work require a different skill set than ones needed in other lines of audio production work. Live sound engineers must be technically competent and skilled at troubleshooting. Their job entails making countless connections among the many pieces of audio equipment in a setup, and in a touring system this means that connections will have to be made and unmade each evening, night after night. A grasp of signal flow concepts, impedance, and level matching, plus the ability to quickly remedy problems, is critical to success in live sound.

Another distinctive feature of sound reinforcement is that live sound mixing is a constantly unfolding activity, changing moment by moment, unlike the recording studio where one can plan and rehearse a mix until perfectly executed. Occasional mixing errors in live sound tend to be quickly forgotten, but in a recording studio mistakes in the final mix will live on forever. Recording engineers can seldom avoid those "I wish I had fixed that" mixes at least once in their careers.

Compared to the audio field's other branches, live sound reinforcement's history is a short one; the specialty has grown quickly. Early **PA**, or public address, systems were just a microphone for the vocalist, a small amplifier, and one or two loudspeakers. Even after the high-sound-pressure-level "British Invasion" ushered in a revolution in the sound industry, reinforcement systems remained rudimentary by present standards. Films of the Beatles' historic performance at Shea Stadium show a stage setup devoid of **foldback monitors**, having only a handful of microphones and speakers. Live sound reinforcement has grown from these early beginnings into a full-fledged industry.

Sound reinforcement systems range in size and complexity—from a single microphone connected to an amplifier/speaker combination to the giant systems having scores of microphones, direct boxes, and other inputs to multiple automated mixing consoles, coupled with amplifiers capable of tens of thousands of watts driving dozens of high-efficiency speaker enclosures, and elaborate monitoring systems. Within the live sound industry, the term *public address* has come to refer the kind of sound systems found at sports stadiums, racetracks, retail stores, and other large facilities where paging, announcements, and/or background music are required. **Sound reinforcement** describes systems used for live music, theater, and other stage events, which together can be subdivided into touring systems and fixed or permanently installed systems. Touring sound reinforcement contracting companies range from small local providers with a single small system to giant international companies, such as Clair Brothers and Sound On Stage, that regularly have dozens of massive systems on the road supporting musical acts worldwide. Some big international acts that need days to erect their sophisticated sound, lighting, and staging employ duplicate systems that leapfrog from city to city. Despite the growing complexity of reinforcement work, some touring performers carry their own sound. Fixed installations range in size and sophistication too, from small club systems to complicated automated systems in venues such as Broadway theaters and Las Vegas showrooms.

FEEDBACK CONTROL

As in other audio specialties, certain general principles apply to sound reinforcement systems at all levels of technical sophistication. The principal challenge of live sound is to achieve sufficient gain without producing feedback. Audio feedback is that irritating, possibly ear-splitting squeal that occurs when the output of a loudspeaker is picked up by a microphone, reamplified, then output again, picked up, and reamplified endlessly. This feedback loop creates a self-sustaining audio **oscillation**. The frequency at which an oscillation occurs is determined by the frequency at which overall system gain is greatest and depends on a number of variables in the signal chain. Uncontrolled feedback not only can damage equipment but also can inflict permanent damage on the hearing of listeners within earshot.

Avoiding feedback is not difficult. The location of speakers and microphones is the first and the most critical decision an engineer makes when setting up a sound system—proper choices here are crucial to obtain the greatest gain before feedback. The key point is that loudspeakers must be in *front* of the microphones. Figure 12.1 shows the proper physical placement of sound reinforcement speakers in relation to microphones. Surprisingly, people who are unfamiliar with the causes of feedback often overlook this simple step, and this alone severely limits sound levels that can be achieved before squealing and rumbling starts.

MICROPHONE USAGE

Microphone selection and placement is another vital element in setting up live sound-reinforcement systems. A few engineers refer to their collection of microphones as "paints on the artist's palette." Each microphone design offers certain

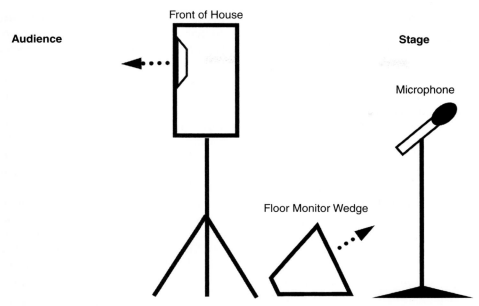

FIGURE 12.1   Correct placement of microphones and speakers in a sound reinforcement setup requires that speakers' sound is directed away from open microphones.

advantages and disadvantages for a given usage. The physical design of the microphone, irregularities in its frequency response, and its polar pattern are factors that should be weighed in selecting a microphone for a specific instrument or voice in a particular venue. In order to minimize pickup from off-axis sources, nearly all microphones chosen for live sound are **cardioid** or **unidirectional**. In fact, engineers sometimes utilize **super-cardioid** or **hypercardioid microphones** to obtain the greatest rejection of off-axis sound. The one-sided directional polar pattern of the cardioid microphone discourages pickup of sound from the main speaker array and the foldback monitors. The **null**, or the side of the microphone offering the greatest rejection, is nominally 180 degrees from its front. The microphone should be positioned so that the null faces the monitor, in order to allow higher monitor levels. There is a trade-off in using this microphone: performers must stay within the narrower on-axis region of the microphone's pickup pattern or else their levels will be erratic.

TUNING A SOUND SYSTEM

Tuning a sound system to the venue is necessary to achieve optimum equalization. Each venue will have its own unique acoustic conditions that color sounds emanating from loudspeakers. In indoor venues these characteristics may even contribute to standing waves that cause room resonances and **room modes**. Room modes result in certain frequencies that are suppressed or certain frequencies that resonate, making them seem louder than others. In large spaces this is often a problem with frequencies in the lower bass range.

Highly experienced live audio engineers may be able to tune their system using a familiar piece of music played over the system. Others are advised to use alternative methods. Live sound systems often make use of precision one-third-octave equalizers between the console's main outputs and amplifiers driving the main speaker array. The one-third-octave equalizer's narrow frequency bands afford excellent control over troublesome resonances. Equalizers like this can be used in a simple method of tuning a system to reduce feedback tendencies. As microphones are activated and their gain increased, oscillation at certain frequencies will begin before others. These frequencies are related to the room modes and peculiarities of system components. Suspect frequencies can be deemphasized by reducing their gain through the equalizer, either by trial and error or by recognition of the range at which ringing is occurring. Some manufacturers offer equalizers that show operators which frequencies exhibit excessive amplitude. On faders of equalizers like this, peak indicators flash to alert engineers that certain frequency bands are nearing oscillation and should be attenuated.

Engineers who are more exacting might prefer a scientific approach, one that assures them that the acoustic output of the system is truly **flat**. Equalization adjustments made while feeding white noise through the system can achieve this with the aid of a **real-time analyzer (RTA)**. White noise is a signal containing equal levels of all frequencies across the audio spectrum. In this method, a calibrated microphone is positioned at the mix engineer's position and connected to the input of a real-time analyzer. The microphone receives the white noise through the main speakers, and the analyzer visually displays the range of frequencies being picked up by the microphone. It is then merely a matter of applying adjustments to a graphic equalizer connected between the console output and the amplifier input so that levels at all frequencies are the same. Many versions of RTA are now available as software programs for laptop computers. Figure 12.2 contains a screen shot of a software-based real-time analyzer on which the right channel shows a weaker high-frequency response.

A new tool for live sound reinforcement is the outboard graphic equalizer with integrated, or automatic, **feedback controller**. The feedback controller combines the features of a real-time analyzer with a graphic equalizer and compressor/limiter. The RTA detects developing oscillations at specific frequencies by sensing increased levels at its line input. A microprocessor calculates which frequency band is increasing in amplitude and passes the result to the compressor/limiter section of the controller. This section, in turn, instantly reduces the gain in the frequency bands where the incipient oscillation has been detected. Not all live sound specialists accept this tool, however. Some veteran engineers remain skeptical of the feedback controller's capacity to manage feedback under actual live-performance conditions.

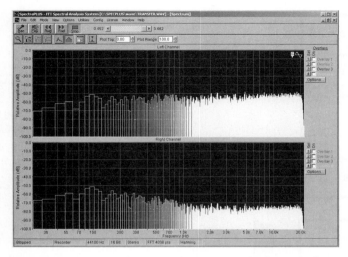

FIGURE 12.2   A real-time analyzer screen display provides a visual representation of the frequency response of an audio signal. In this example the right channel at the bottom exhibits higher bass and lower high-frequency response than the left channel.

LOUDSPEAKERS AND MONITORS IN LIVE SOUND SYSTEMS

Most compact club-sound systems make use of full-range speaker enclosures that have built-in crossover networks. As detailed in Chapter 10, Monitors and Loudspeakers, a crossover network divides the full-range signal from the amplifier into separate frequency bands for each of the different drivers in the enclosure. Crossovers may be two-way (high- and low-frequency outputs), three-way (high-, mid-, and low-frequency outputs), four-way (low-, low mid-range, high mid-range, and high-frequency outputs) or may even have more **crossover points**. In live-sound reinforcement, these networks are vital for protection of high-frequency drivers, which could easily be destroyed if a low-frequency signal were applied. In larger touring systems, crossovers are frequently not contained with speaker enclosures but are external. In such systems the crossover is inserted between the console's output and the amplifier's inputs. Crossover points can be manually set by the engineer and the outputs routed to the individual amplifiers that feed speaker enclosures designed for different frequency ranges. Figure 12.3 illustrates how a full-range signal from the **front-of-house** (**FOH**) console is split by the crossover, and then routed to the separate amplifiers for the high-, mid-, and low-frequency speaker enclosures. In live sound, front-of-house refers to the portion of the venue accessible to audiences and not areas used by performers, such as the stage, dais, or performance platform.

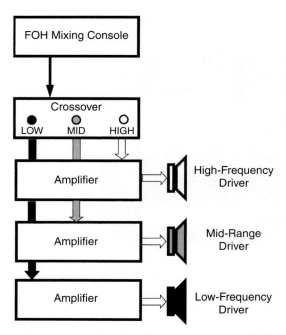

FIGURE 12.3 Outboard crossovers are used in large, live-sound reinforcement systems to divide a full-range signal into separate bands that are routed to individual amplifiers and their associated loudspeakers.

The operation of even medium-sized sound systems really means managing several separate mixes simultaneously. The audience will hear the main mix, often referred to as the front-of-house (FOH) or simply the house mix. Stage performers usually receive a separate mix that they hear on a completely different loudspeaker system called foldback monitors. There may even be several different monitor mixes, each tailored to the preferences of individual musicians, sent to different foldback loudspeakers. For example, the vocalist may want to hear a mix heavy on vocals, with just enough of other performers to keep from getting lost in the song, or the drummer may want to hear only the bass and rhythm guitars in order to keep accurate time.

Most live-sound consoles allow the monitor mix to be set up either as pre-fader or post-fader, meaning that the signal sent to the monitors can be adjusted either before the console channel faders or after the faders. Figure 12.4 shows the difference between pre-fader and post-fader monitor feeds. Of course, pre-fader mixes eliminate the action of faders in establishing an overall mix. Setting up the monitor mix as pre-fader has the advantage of leaving the artists' mixes unaffected when the engineer changes the front-of-house mix. Once pre-fader foldback mixes have been set to the artists' satisfaction, they should require little further attention during performance.

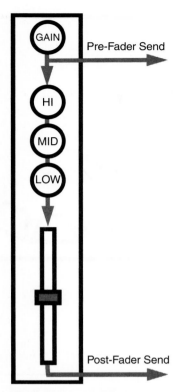

FIGURE 12.4   The choice of pre-fader or post-fader monitor feeds determines whether the fader and equalizer control sound levels sent to the monitor.

In the largest and most complex live sound-reinforcement events, the system setup is likely to be a bit different, with a separate monitor console just off-stage. In this arrangement, a microphone splitter is required to divide the signal from each microphone, sending one signal from each mic to the front-of-house mix and another to the monitor mix. Typically, splitters are built into the stage box of the **snake**, although otherwise the signal may be split in the monitor-mixing console. Snakes are prepackaged multiple runs of cable for microphones or other pieces of equipment. Snakes are more durable than individual cable runs and they are far less likely to result in cable snarls. Figure 12.5 illustrates the signal flow in a live sound system. The inset at the top of drawing shows the function of the microphone splitter in the stage box. In the scenario presented in this figure, the monitor engineer's only task is to manage the separate monitor mixes required by the musicians. Monitor consoles permit engineers to listen to any of the mix outputs. Some monitor engineers use headphones to check monitor mixes, but others favor monitor speakers installed at their mixer's location.

An alternative to the classic front line of monitor **wedges** used in stage monitoring has been gaining in popularity—the in-ear monitor. These tiny earphones,

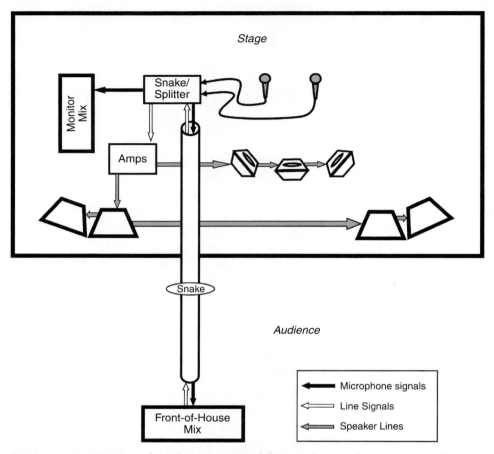

FIGURE 12.5   Live sound reinforcement signal flow can be complex. Here microphone signals are split at the stage snake and routed to both the monitor and FOH mixers. The output signal from the FOH mixer then travels back through the snake to amplifiers driving the speakers.

often referred to as **ear buds**, are placed into the outer ear canal. Ear buds range from inexpensive off-the-shelf generic models not unlike those included with pocket radios and MP3 players to custom-made models that are molded to the exact shape of an individual's ears. Figure 12.6 is a photograph of professional ear buds for use with in-ear monitor systems. To free the performer to move about the stage away from monitor wedges, wireless in-ear monitors can be purchased. Wireless in-ear systems employ small transmitters that receive a monitor mix from the console. A small body-pack receiver to which the ear buds are connected picks up signals from the transmitter.

To protect the hearing of users, in-ear monitor systems normally include a limiter in the signal path. The limiter is set to instantly compress all abnormally

FIGURE 12.6   In-ear monitors

high signals arriving from the console. In the event of a sudden spike in amplitude or runaway feedback, the sound will be clamped at a safe level. Of course, feedback is much less likely when onstage monitor speakers are not used. Ear buds should seal the user's ears completely; thus, ambient sounds will be greatly attenuated. As a consequence, some performers may request that a room microphone be included in their mix so that they can hear crowd reactions. This also gives them a natural sense of space that will be lacking from the monitor signal from the console.

PREPERFORMANCE SETUP

Regardless of the type of performance, the kind of venue, or the scale of the sound system, most live sound engineers insist on a **sound check** prior to the event. The sound check is the time to set levels, equalization, and effects. In addition to tuning the system, the sound check affords the engineer an opportunity to test each cable and microphone, to confirm the proper routing of signals through outboard equipment, and to examine the operation of every single item of equipment.

Once the sound reinforcement system is tuned and its output flattened for the room, the process of establishing proper gain levels for each input will be the next step. Setting levels for a musical performance can be a tedious business in which the engineer is treated to several minutes of repetitive notes from each drum in the drum kit, and each instrument in the group. Getting levels right for drums and other percussion instruments represents one of the biggest challenges. Live sound engineers frequently complain that drummers will sometimes "sandbag" during the sound check; that is, they will tap lethargically on the drum set during a sound check, only to attack the drums viciously during the performance. In an effort to prevent that problem, the engineer may establish a gain level for each microphone and then ask the drummer to play the kit as it will be played during the performance. Other instruments' outputs can certainly fluctuate during a performance too. Guitarists will boost their volumes during solos, and electronic keyboardists will often do the same. This instinctive action by performers will undoubtedly require some compensation by the engineer. Live sound engineers tend to agree that if all performers kept their instrument levels consistent, mixing the show would be much easier. This, of course, ignores the natural changes in level that good musicianship demands. Live mixing is additionally made more difficult by the fact that performances and performers are active, responsive, and spontaneous; hence, somewhat unpredictable. Inevitably, musicians' stage levels creep up during a show, and microphone levels at the close of the show will rarely be the same as those set during the sound check.

Automated mixing consoles have been steadily making their way into live performance. Even though traditionalists at first thought them too expensive, complicated, and delicate for road use, those concerns have diminished, at least for national acts and the more ambitious regional acts. Permanent live sound systems are more apt to make use of automated consoles. There are two great advantages to the use of automated consoles in mixing the same performers night after night. First, they always offer an established starting point. Rather than starting from scratch each night, a default setup can be worked out, stored, and later recalled. Gain settings, fader levels, equalization settings, and routings can each be stored and retrieved at the start of each sound check and each performance. Second, for complicated shows requiring set changes, changes in instrumentation, multiple groups of performers, or other events demanding quick changes, each setup change can be stored as a scene and instantly recalled when needed. In this way, individual levels for a three-band show can be determined, saved during the sound check, and then recalled as the show moves from act to act. Beyond this, the automated mixer should be treated just like a conventional mixer because engineers rarely, if ever, make use of automated mixing functions of the console during actual performances.

LEVEL MANAGEMENT

Here is a caution about activating a system: engineers must turn *on* amplifiers *last,* and in shutting a system down, amplifiers must be turned *off first.* If amplifiers are powered up when the mixing console is switched on or off, the resulting loud bang will probably damage the loudspeakers.

Increasingly, venues require live sound engineers to monitor sound pressure levels of performances and to hold them within specified limits. As awareness of the risks of sustained exposure to high sound pressure levels has spread, many in the entertainment industry have taken measures to ensure safe levels. This has been partly spurred by a trend toward municipalities' enacting noise ordinances aimed at limiting SPL (sound pressure level) from all sources. Also, venue managers worry about the potential liability for damage to patrons' hearing. Outdoor venues in close proximity to residential neighborhoods are especially likely to set SPL restrictions, and even indoor performances have recently begun to face limitations. As a result, a stock tool for engineers nowadays is a sound pressure level meter kept at the house mix position.

LIVE RECORDING PROCEDURES

Technological innovations have radically altered the practice of making live concert recordings, and they have shortened the turnaround time needed to deliver recordings to consumers. Previously, live recordings were made either by sending a feed from the FOH mixer to a stereo recorder, or by splitting signals from each microphone at the stage (some snakes can provide a three-way split) and then routing signals to an isolated console. Usually this console would be housed in a truck outside the performance site. At the remote console, an engineer would create a multitrack recording of the performance for later mixing in a studio. Some concert recordings were also produced through the use of a stereo microphone array located at an optimum spot in the venue and by capturing the house mix and crowd sounds.

    Each of these conventional strategies has shortcomings. Recording a stereo feed from the FOH console may yield a mix that is not well balanced, often with vocals much too high in the mix. Level adjustments made by the FOH engineer in response to stage conditions will probably be distractingly apparent in the recording. Moreover, recordings made by capturing the house mix usually lack clarity and definition, with little chance for a postproduction fix. Finally, multitrack recordings produced from split microphone outputs are expensive.

    Recently, a completely different technique for live concert recordings has become popular. In this method, a stereo mix is created from the pre-fader auxiliary sends on each channel of the FOH console. That stereo mix is routed to a real-time CD recorder. As soon as the disk is written, it can either be mastered—equalized and edited with levels optimized—or placed directly into a bank of disk duplicators. This approach allows artists to offer live recordings immediately after the concert's conclusion. Using high-speed computers, disk writers, and printers, CDs can be packaged and ready for sale to attendees as they exit the performance site.

TROUBLESHOOTING GROUND LOOPS

A problem that frequently plagues live sound engineers is the presence of **hum** in signals. Hum can be introduced into the audio in a number of ways, but it is usually related to poor cable shielding, dimmer hash, or to ground loops.

    Professional live sound reinforcement systems use **balanced** line cabling to connect the console with outboard devices, crossovers and equalizers, and amplifiers.

Balanced cables incorporate three conductors: hot (+), cold (−), and shield. These cables are terminated with **XLR** or quarter-inch tip-ring-sleeve (**TRS**) connectors. In a balanced circuit, hum and noise can be led away via the shield to the **earth** or **ground**. Unbalanced cables have only two conductors, and the cold (−) and shield must share a common conductor. Consequently, unbalanced cables are more susceptible to hum and noise.

The importance of proper grounding can hardly be overstated. Many experts advise the use of a single-point grounding scheme, in which all items of equipment are attached so that there is only one path to a common ground. When sound system components are inadvertently plugged into circuits that have different paths to the ground, the result can be a hum-producing difference in potential, or what is commonly called a ground loop. When in doubt about grounding, the safest method of powering a sound system is to plug the console into the same power circuit as the amplifiers. Consoles do not draw a heavy current, and normally can be powered safely from the same circuit as high-power amplifiers.

AUDIO CONNECTORS

Because live sound personnel usually set up their own equipment, the following discussion addresses some of the most common connections and cabling protocols used for professional audio. Regrettably, audio professionals have been slow to adopt uniform standards for cabling and connectors. Although de facto standards have begun to emerge, they have not been articulated by an official agency; only the **Audio Engineering Society** (AES) has published specifications for a few connector types. As a result of this general disinclination to impose standards, audio studios and live sound systems commonly use a broad array of connectors and cable types.

Perhaps the most familiar connector in professional audio is the three-pin XLR type developed by Cannon (later ITT Cannon). It is a near-universal standard for low-impedance microphones, balanced lines, and AES/EBU digital audio connections. Cannon originally marketed the P series connector and somewhat later the UA series, both of which were XLR antecedents. The P and UA connectors had threaded collars to lock the connector in place. Subsequently, Cannon introduced a smaller X connector that enjoyed some degree of acceptance in audio use. Cannon then added a latching mechanism to the X connector and set the contacts in a flexible rubber compound, labeling it the XLR. This is now a generic term for the 3-pin latching connector used throughout all audio fields. The photograph in Figure 12.7 shows male and female types of the XLR connector. For low-impedance microphones the AES recommends that pin 1 = shield/ground, pin 2 = hot (+), and pin 3 = cold (−).

Another connector widely adopted in audio is the **quarter-inch connector** or longframe connector, developed by Western Electric-AT&T. The quarter-inch refers to the diameter of the connector shaft. The quarter-inch connector is still occasionally referred to as the phone plug, as it was once the standard connector

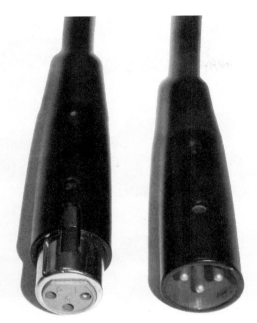

FIGURE 12.7   Female (left) and male (right) XLR connectors used for many professional audio connections.

used in telephone central office switching equipment. The quarter-inch connector is manufactured in both unbalanced/monaural and balanced/stereo versions. The mono version is called the tip-sleeve (**TS**) configuration, while the stereo version is the tip-ring-sleeve (TRS) design. In Figure 12.8, the quarter-inch connector on the bottom is a monaural tip-sleeve version, the middle connector is a stereo or balanced tip-ring-sleeve version, and the top connector is an eighth-inch tip-ring-sleeve connector. The quarter-inch phone plug and jack are used for connections between line level equipment, headphones, patch bays, musical instruments, and, infrequently, for microphone signals.

A smaller version of the quarter-inch phone device is the bantam or **TT** (tiny telephone) connector. Developed by Switchcraft, Inc., the TT connector is slightly smaller in diameter and length. It is used in many audio patch bays. The Radio Corporation of America developed the familiar **RCA connector**. RCA jacks, as shown in Figure 12.9, are also called phono plugs, and less commonly cinch connectors, or AV jacks. They are used for composite video and audio interconnections on videocassette recorders, and line level audio for home stereo receivers. As the term phono plug suggests, RCA connectors have long been used to connect phonograph turntables to preamps and receivers. The RCA connectors are not so common in professional audio applications, but they do serve as the standard connector for SPDIF (Sony/Philips digital interconnect format) devices as found on digital player/recorders.

FIGURE 12.8   Phone plug connectors from top to bottom: Eighth-inch stereo mini, quarter-inch tip-ring-sleeve, and quarter-inch tip-sleeve.

One of the smallest audio connectors is the **eighth-inch mini plug**. Its diminutive eighth-inch-diameter barrel makes it less rugged and rather difficult to repair, so it has not enjoyed much acceptance among audio professionals. Nonetheless, it is common on computer sound cards and consumer audio equipment. Like the quarter-inch connector, eighth-inch mini connectors are manufactured in both monaural/tip-sleeve and stereo/tip-ring-sleeve versions.

Speaker connections often use **binding posts**—sometimes called *five-way* binding posts—that accept **banana plugs,** spade lugs, or bare wire. This appears to be the preferred connector for permanently installed studio speaker monitors, but they are sometimes found on older sound reinforcement speakers and amplifier outputs. Many live sound speaker connections, especially in smaller systems, use quarter-inch connectors. Large high-quality live sound speakers and amplifiers increasingly make use of the **Speakon** connector, developed by **Neutrik**.

FIGURE 12.9   Male RCA phono plug.

Speakon connectors are available in two-conductor (NL2) to eight-conductor (NL8) configurations. Speakons are rated at 250 V and feature a latching twist lock design that prevents accidental disconnections. Figure 12.10 shows a two-conductor type of Speakon connector on the left.

MIDI equipment uses a 5-pin **DIN** connector of the sort pictured on the right in Figure 12.10. In most cases, pin 2 is attached to the shield, pin 4 carries +5 VDC from the MIDI output, and pin 5 carries the MIDI signal. Some manufacturers use the spare pins for additional purposes.

The **BNC** connector was developed for use by radio frequency signals, and it is only infrequently encountered in audio settings. The BNC is a twist-and-lock connector for coaxial cable developed in the 1940s. Although it is primarily used on video devices, BNC jacks can be found on audio data interfaces, including time code or word clock signals applied to synchronize various pieces of audio equipment such as automated consoles and digital tape machines. Other connectors used for digital equipment an audio engineer might encounter include D-sub serial connectors, RJ-45 Ethernet connectors, optical (fiber optic) connectors, and dozens of others.

Cables found in audio applications tend to be of three types: one-conductor-shielded, two-conductor-shielded, and unshielded. Some call the single-conductor shielded cable instrument cable, because it is used for electric guitars, keyboards,

FIGURE 12.10   On the left is a Speakon connector used to connect speakers to amplifiers or to other speakers. On the right is a five-pin DIN or MIDI plug used for interconnecting musical instruments, computers, and similar devices.

and other electronic music instruments. Single-conductor shielded cable is appropriate for line level audio runs, for instance in connecting outboard devices, or connecting a mixer's output to an amplifier's input. It consists of braided or spiral-wound stranded wire wrapped around a single insulated conductor.

Two-conductor shielded cable is also called balanced or microphone cable. This type of cable can be employed in low-impedance microphones, for balanced line level connections, and for situations where two signals are present in a single cable, such as in mixer inserts where sends and returns are present in the same cable. Balanced lines allow longer cable runs without pickup of noise or hum. Figure 12.11 shows three types of double-conductor shielded cables found in audio applications.

Unshielded cable is suitable *only* for speaker connections; the use of unshielded cables for line level signals is certain to result in noise problems. It is imperative to use cable of sufficient size for loudspeakers. Wire gauges are standardized in the United States by a system known as the American Wire Gauge (AWG), formerly called the Brown and Sharpe system. In this system, there is an inverse relationship between AWG size and conductor diameter. Gauge numbers range in value from 0000 (called *four ought*) to 50, with 50 being the thinnest. Telephones use AWG 20, 22, or 24 conductors and household wiring may be 12 and 14 gauge. Larger-diameter wires are required for higher currents and larger-diameter wires will result in less voltage drop over long distances.

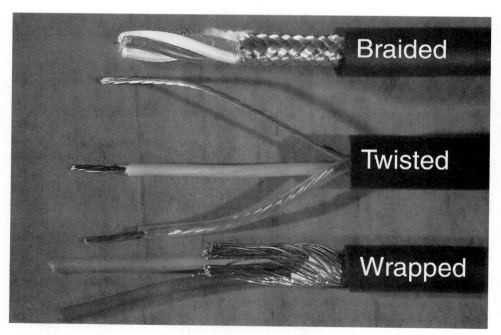

FIGURE 12.11   Three types of shielded cable used in connecting microphones and balanced line level devices.

The match of an amplifier's output impedance to its load will affect its efficiency and reliability. Smaller-diameter wire has greater resistance, a factor that may drastically affect the load an amplifier sees at its output. Equipment manufacturers suggest using wire no smaller than 16 AWG for speaker connections on short runs, but urge the use of larger wire for longer runs. At any rate, there is no penalty for using wire larger than necessary for a run, but there surely will be for using wire smaller than needed.

Finally, here are a few words on the conductivity of conductors used in cables. On the whole, conductors are manufactured using metals that have the least resistance possible. By far the most popular metal for use in wire and cable is copper because it has good conductivity. Aluminum is sometimes used, but its resistance is slightly higher than copper. Unfortunately, other metals with low resistance or high conductivity are silver and gold, in that order. These precious metals are much too expensive for use in wire and cable. Gold and silver connectors are being manufactured, however, and even though they are costly they provide the best connection between cable and equipment.

LOAD IMPEDANCE MATCHING

As indicated at the beginning of this chapter, audio professionals tend to draw a clear distinction between sound reinforcement and public address systems. Sound reinforcement refers to concert or live theatrical sound, and public address mainly refers to announcements or paging. While a sound reinforcement system may have a host of inputs for voices and instruments, it has comparatively few output loudspeaker locations. In PA systems the reverse is true; they have a small number of inputs, perhaps a single microphone, but a large number of loudspeaker locations. Public address systems may include a great many speakers scattered over a widespread area.

The long cable runs needed in PA systems pose a problem in matching amplifier output impedance. Amplifiers are normally intended to match a line impedance of 4 ohms or 8 ohms. As the system grows larger, the total resistance of the cable run will become higher, thus causing the impedance to increase proportionally. When this happens line loss can rise to unacceptable levels.

If the number of required speakers is large, the problem can be compounded. When speakers are simply wired in parallel, their combined load impedance will be less than their individual values. This can lower the overall load impedance to levels that so badly mismatch an amplifier's output impedance there will be a danger of amplifier damage. Connecting speakers in a series configuration can raise the total line impedance, but this approach also has a serious drawback: if one speaker fails, all speakers in the series will lose power. In some systems, it may be possible to wire speakers in series and parallel combinations so that the total line impedance remains within tolerance limits. This can become a complicated solution. Figure 12.12 illustrates several possible wiring configurations for a sound system containing multiple loudspeakers.

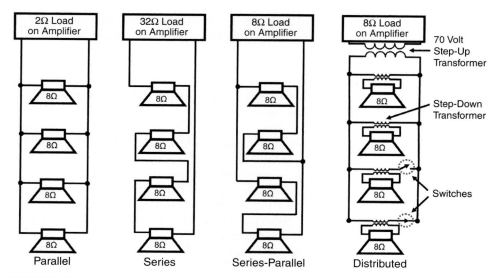

FIGURE 12.12   A sampling of arrangements used to manage differing load impedances in multiple loudspeaker systems.

The solution to these problems for most large PA systems is to use a **distributed sound system.** Distributed sound systems use transformers to step up the amplifier's output—usually to 70 V but sometimes to 25 or 100 V—before sending audio to the speakers. Each speaker is outfitted with an appropriate line-matching step-down transformer to lower the line impedance and voltage to match the speaker. This approach has several advantages, chiefly that power losses will be significantly reduced at the higher voltage. Distributed loudspeakers can be switched on and off or divided and routed with little concern for impedance matching. Although electrical codes do not require conduit for runs of 70 V or lower, signals should be treated with the same care as 120 V household current because they carry much the same risk of electrical shock.

PA amplifiers used in distributed systems are often equipped with 25 V and 70 V outputs as standard options. PA horns for use in outdoor settings such as stadiums or racetracks and speakers designed for installation in suspended ceilings can be purchased with step-down transformers already built in. It is not uncommon for these speaker transformers to include multiple input taps to match different line voltages.

Distributed sound techniques enable PA systems to cover immense areas, composed of literally hundreds of loudspeakers and wire runs of several miles. In setting up these systems, care must be exercised to ensure that the amplifier's rated output is equal to or greater than the total power required by the combined speaker load. If there are ten speakers, each expected to deliver 10 W of sound power, an amplifier rated for at least 100 W output is a must. At one time, the audio quality of distributed systems was considered too low for music use, but modern transformer designs make them appropriate for all types of audio.

A FINAL OBSERVATION

It is still true that live sound engineers who work for touring shows require a strong back! Until the day that they become respected as professionals having special knowledge and talent, they will not only be counted on to operate sound equipment but they will generally have to unload, set up, tear down and reload their own gear. Live sound mixing requires quick reflexes, anticipation, technical expertise, and inventive thinking when problems crop up—as they inevitably will. In no other type of audio production is Murphy's Law more likely to come into play. It seems that when something can go wrong, it will, and at the worst possible moment. Engineers sitting behind the console when a critical component fails or screeching feedback howls find that the stares of hundreds or thousands of listeners can be very disconcerting (no pun intended).

CHAPTER THIRTEEN

RADIO STATION OPERATIONS

This chapter describes equipment configurations and audio production practices found in radio broadcasting. While many of the tools and techniques are common to other applications, radio does employ some specialized hardware and software, and unlike other portions of the audio field may require operators to be both technician and performer. Radio production is a business of split-second timing, and content is intensely live, continuous, and transient.

RADIO'S ORIGINS

Broadcasting was born in 1920 as an audio-only medium. Previously in its short life, radio transmissions had been mainly used for point-to-point communications by government, sea transport, and amateur experimenters. In broadcasting's earliest years, audiences were mostly hobbyists who enjoyed listening to voices and music captured from the "ether" (as it was then called). Much of the thrill for listeners lay in receiving radio signals from distant locations, a practice known to hobbyists as "DX-ing"; hobbyists themselves were known as "DX-ers." These listeners only had the use of simple receiving setups such as crystal radios and wire antennas. Nevertheless, broadcasting opened radio to the masses.

As stations began to offer more attractive programming, radio grew rapidly in popularity, with audiences less concerned about technical aspects of receiving signals. Broadcasters became more sophisticated in their audio production techniques and, with improved and more appealing shows and performers, radio began to rival the popularity of motion pictures. Indeed, many radio performers rose to become stars equal in magnitude to those in Hollywood.

By the time of radio's so-called Golden Age in the 1930s, studios had evolved from the rooftop tent studio used by KDKA in Pittsburgh for its inaugural 1920 broadcast to large studios that could accommodate both orchestras and large studio audiences. Over the years, radio broadcasters have had to reposition themselves in response to the emergence of new media and to marketplace changes. In this way, radio has maintained its importance despite competition from television,

VCRs, DVDs, and the Internet, as well as the fickleness of audiences' programming preferences. Meanwhile, station facilities continued to change with the times, reflecting the shift from lavish live programs to disk jockey shows featuring musical recordings, and from drama and comedy to talk and information. Across the years, and despite shifts in programming, the demand for good radio production techniques never declined.

THE ON-AIR CONTROL ROOM/STUDIO

To match programming needs over the years, the relationship between radio on-air studios and their control rooms has changed greatly. During radio's Golden Age, on-air studios where actors, singers, musicians, and announcers performed were separate from control rooms where technical staff operated the audio console and related equipment. It was up to the studio engineers, commonly known now as board operators or "board-ops," to ensure that audio sources such as microphones were opened as needed and that good audio levels were maintained throughout broadcasts.

When television began to offer the same variety of entertainment programming—with pictures—in the late 1940s and early 1950s, radio audiences began dropping sharply. To combat the trend broadcasters were forced to alter programming. By the end of the 1950s, radio had largely abandoned live drama and musical programs in favor of a mix of programs of recorded music hosted by disk jockeys and newscasts, often with a measure of sports programming tossed in.

Radio stations reduced the number of people on both their programming and engineering staffs, and this led to "combo operators," disk jockeys who operate their own audio consoles. Combo operators not only managed mixing and audio levels, they cued records, and started turntables and other studio equipment as needed, all the while performing as on-air talent. There were a few exceptions to this pattern, mostly at stations and networks with strong union contracts, where studio engineers handled the technical duties and the disk jockeys were responsible only for patter between music and commercials.

With little need to accommodate large groups in the studio and with the rise of combo operators, most radio stations began to use a single room as both control room and on-air studio. A typical radio control room can be seen in Figure 13.1. Station staff usually referred to the room simply as the control room, or sometimes as the **air studio**. Typically, a second, smaller studio filled the role of a news booth where the station's news staff could read newscasts. At smaller stations, reading news on the air was just one of several additional duties of the combo operator.

Because radio schedules relied less on live programming, stations seldom made provision for studios separate from their control rooms. However, in the 1990s talk radio became an important format and this meant a new division of duties among talk show host, producer/board-op, and call screener. This, in turn, led stations to reconfigure facilities so that air studios became separate from their

FIGURE 13.1   A contemporary radio station control room is designed for a single person to serve as both on-air talent and equipment operator. In this installation, an operator faces the mixer with playback devices to the left and right and a computer to the far left.

control rooms again, but on a much smaller scale. Likewise, in some major markets on-air personalities—a term that has largely supplanted disk jockey—rely on an engineer/board-op to coordinate show elements, thus permitting on-air talent to focus on their performance.

SPECIAL FEATURES OF THE RADIO CONSOLE

In radio, as elsewhere, audio mixers are known by a variety of other names, such as the console or board. For decades, radio consoles were manufactured with rotary faders that an operator rotated to adjust the gain of selected audio sources, much like a conventional volume control. Figure 13.2 shows a small rotary fader radio mixer. Today, a radio board operator is more likely to work with consoles having linear faders similar to those seen in the recording-studio field.

Consequently, radio station consoles have much in common with those used in settings such as multitrack studios and sound reinforcement, but there are significant differences that reflect the particular requirements of radio production. Possibly the most striking difference is the addition of the cue feature as described in Chapter 5, Mixers. In radio, *cue* has a different meaning than the one applied to consoles intended for recording or live sound reinforcement. With the cue feature,

FIGURE 13.2   A small rotary potentiometer radio mixer.

a radio board operator can preview audio material off-air, so as not to interrupt programs being broadcast. The cue feature permits an operator to locate the beginning of a recording and thus ready it for instant on-air playback. With the exception of microphone inputs, most channels on the radio console are equipped with the cue feature. Cue is activated by pulling the fader slightly beyond the off position, past a detent in the fader track. A slight click will be observed when placing the fader in the cue position. In Figure 13.3, two console channels with the cue function are depicted, with the left channel shown with the cue function activated. An alternative arrangement is to provide a specific cue button for each nonmicrophone channel, enabling an operator to access the cue feature without moving the fader. This is generally less favored by operators because it tends to increase the risk of accidentally cueing material on-air.

In most control rooms, the cue amplifier and speaker system is not intended to provide high-quality sound. If it is necessary for an operator to judge the audio quality of a source off-air, this can be done by routing the signal to the audition bus by means of another selector associated with each fader. Most radio consoles provide at least selectors for a program bus and an audition bus on each input channel. More complex designs offer additional selectors on each input module, making it possible to route signals to other output buses. These outputs may include auxiliary, utility, and mono buses used to send audio to recording equipment, telephone interfaces, two-way radio, and other devices. The number of output buses determines a mixer's ability to connect with those external items of equipment. Figure 13.4 illustrates an example of signal flow options for a single channel in a radio console.

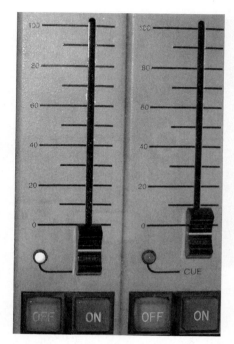

FIGURE 13.3   The linear fader on the right has been faded all the way to zero while the fader on the left has been pushed past the zero position so that it has snapped into the c position. Note the illuminated cue LED on the left fader.

When a program output bus is selected as a source and its fader is up, its audio is mixed with audio from other program sources. The output of the console's program bus (or buses, if stereo) is then routed through a series of audio limiters and compressors, to the transmitter through telephone lines or a microwave link, and finally to the antenna. In the case of Web radio, audio signals are sent from the console to a computer for conversion to a streaming format such as Windows Media or Real Media, then to a server for distribution via the Internet. Figure 13.5 illustrates an example of the various sources and destinations that may be connected to a radio console.

By selecting audition both for a given audio source and the studio monitors or headphones, an operator can hear the audio off-air. This is often done to judge whether its quality measures up to broadcast standards. In preparing for a **remote** broadcast such as from a football field or car dealership, a board operator might listen to the remote connection via the audition bus to assess noise levels and the audio bandwidth of the telephone line or radio link. Older consoles often designated remote sources by the term *nemo.*

The audition channel of most radio boards can also be used to feed audio material to recording equipment for later broadcast. Material such as weather forecasts, telephone interviews, music requests, or news and sports reports are often

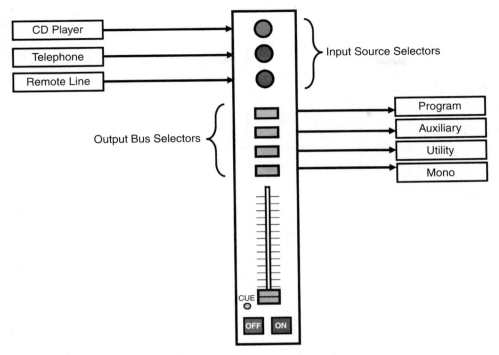

FIGURE 13.4   Signal flow options for a radio mixer channel.

prerecorded to give operators more flexibility in incorporating items into programs. To enhance audio prior to recording, a utility bus might also be used to feed audio from selected input modules to signal processors or other devices.

Utility or *auxiliary buses* are sometimes used in conjunction with telephone interfaces such as those manufactured by Telos systems to feed audio from studio microphones directly to a studio telephone. This allows an operator or program host to speak on the telephone without having to hold a telephone handset. The arrangement also allows others in the studio to listen to callers via the cue channel.

To position a source in the stereo field some radio consoles include pan pots for each input module much like those found on multitrack consoles. This is, however, less common on broadcast consoles. Also, there are usually two or more source selector switches or buttons associated with each channel, allowing the operator to choose inputs to the module. Input source selectors are mutually exclusive—only one source may be selected at a time—but output selectors may normally be switched in any combination.

There might also be a button to select between stereo and mono for each individual source. This control can be very helpful when the operator needs to broadcast audio from a monaural source that has been connected only with the left or right channel of the input module, or from a recording that uses only one

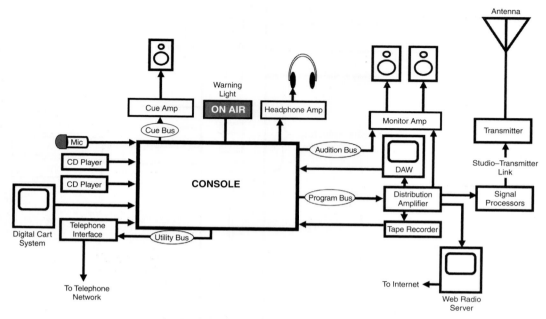

FIGURE 13.5   This block diagram presents a complete broadcast radio station layout, showing signal flow from audio sources through to the transmitter's antenna.

of the stereo channels. By pressing a mono button, the operator feeds audio to both left and right channels and raises its overall gain to match that of the rest of the broadcast. Additional controls found on every input channel are the on and off buttons. The on switches of radio mixers are usually connected so that, in addition to sending audio into selected buses, they also start playback from devices such as CD-players, tape decks, digital cartridge systems, and digital recorders.

Most radio consoles these days are configured as stereo boards, with a single fader controlling both left and right channels of a stereo signal. This is different than recording and live sound reinforcement consoles, most of which include only monaural channels. Another difference between broadcast consoles and those used in recording and live sound is the metering setup. The typical radio console lacks signal-level meters for each audio source, having instead VU meters only for the left and right channel output to the transmitter. These are usually labeled *program VU meters.* Sophisticated radio consoles might also provide left and right channel VU meters for audition, a utility channel, and perhaps a single VU meter for monaural output.

For broadcasting purposes, the VU meter—whether a mechanical moving needle, light-emitting diodes, or a plasma display—will be calibrated in both volume units and percentage of modulation. Since broadcasters have to be concerned about modulation percentages of transmitted signals, engineers tend to observe the percentage scale more closely than the VU scale. One hundred percent modulation on the meter corresponds to zero volume units. The portion of the scale above zero VU is normally colored red. Due to the danger of overmodulation, operators must strive to keep maximum audio peaks out of the red. The meter illustrated in Figure 13.6 is shown at approximately 82 percent modulation, a good average reading for broadcasting.

Even with audio limiters and compressors placed between a console and the transmitter for proper level maintenance, studio operators still must set good levels. Compressors cannot sufficiently boost unusually weak signals, and extremely strong audio signals will overload limiters and compressors. The result can be a distorted signal and splatter on AM or excessive bandwidth on FM—and a possible citation levied against the licensee by the Federal Communications Commission. A good operating practice is to keep frequent and recurring audio peaks within an 85 percent to 100 percent range. At this level the station's limiters and compressors can easily accommodate occasional audio peaks that venture into the red portion of the VU scale.

The monitoring section of a console provides pots for adjustment of audio levels heard by the operator. These may be available not only for control room monitors, but also for headphones and the cue speaker. Switches are provided to allow users to choose among the program, audition, and auxiliary buses for the main control room monitor loudspeaker. Headphones are essential for monitoring if the console is to be used in a combo control room. When the control room microphone is activated, the monitor loudspeaker is simultaneously muted to prevent feedback. Operators can continue to monitor the mixer output by headphones even when the control room microphone is switched on.

FIGURE 13.6   In broadcasting, audio levels should be maintained at an average of about 80 percent on the mixer's VU meter.

MICROPHONES IN THE RADIO STUDIO

The number of microphones installed in a control room studio will be dictated by programming needs. A minimal studio setup might include only one microphone for a single program host, but other studios might provide as many as five or six microphones to serve on-air cohosts, in-studio interviews, roundtable discussion programs, and the like.

High-quality, large-diaphragm cardioid dynamic microphones, such as the Electrovoice RE-20 or the Shure SM-7, are preferred by many producers for radio voice work, although condenser microphones and ribbon microphones are sometimes used as well. Since close mic placement is the usual practice in radio, proximity effects usually associated with cardioid microphones may become an issue. Microphones can reduce or eliminate this effect if they incorporate a bass roll-off switch. The result is an acceptably smooth, flat frequency response at lower frequencies, even if the microphone's distance from the speaker's lips is only a few inches.

Radio studio microphones are customarily attached to spring-loaded desk-mounted booms. These kinds of mounts allow each announcer or on-air guest to easily reposition a microphone as needed. Not only can a user move the extended arm of a mic boom up and down, sideways, closer, or farther away, but the microphone itself may be swiveled up, down, or sideways on the boom. Naturally, heavy microphones require booms with stronger springs to keep them balanced in desired suspended positions. In Figure 13.7, Electro-Voice RE-7 microphones are shown mounted on typical radio control room–type booms.

As they gain experience, announcers and program hosts tend to develop preferences for microphone placement that highlight their voices' character. As a general rule of thumb, it is best to maintain a distance of about five or six inches, approximately the width of a person's hand, between a mic and the speaker's lips. Placing a microphone slightly off to one side with the microphone still pointed directly toward the speaker will minimize the risk of plosives. This placement helps a speaker avoid "popped P sounds" by allowing the burst of air created by certain vocalizations to blow past the microphone rather directly into it. Figure 13.8 illustrates the recommended positioning of a microphone for radio announcers. As a further precaution, a foam windscreen placed over the microphone may be used.

If a microphone picks up excessive breath noise when the announcer inhales, creating a "wheezy" sound, the speaker should move away from the microphone slightly or turn away from the mic when inhaling. Professional performers normally prefer to maintain both a constant mic distance and average speaking level to avoid the possibility of an off-mike effect.

In radio, a microphone that is active is said to be *open*. In a studio, any mic should be treated as open, unless it is perfectly clear that it is not. Many are those who have learned the hard way that a microphone thought to be off was actually open and capable of picking up every embarrassing word uttered. In combo studios, the cue for an open mic is the muting of the control room loudspeaker.

FIGURE 13.7   Spring-loaded booms are commonly used to mount microphones in radio studios. This arrangement brings the mic close to the performer.

FIGURE 13.8   Two views of proper microphone placement, about six inches from the announcer's lips and slightly to the side.

ADDITIONAL CONTROL ROOM EQUIPMENT

Digital audio devices have today supplanted analog equipment at many, perhaps most, radio stations. The digital revolution in radio began with compact disk (CD) players. At first, CDs were mere additions to control room inputs, but eventually they replaced phonograph turntables altogether. Broadcast CD players are more ruggedly constructed than consumer models and incorporate special features that make players more suitable for broadcasting. For example, broadcast CD players include a remote start feature that permits the start of disk playback from the audio console. Upon insertion of a disk, radio station CD players normally cue to the beginning of a track and enter pause mode automatically. A rotary control or an up-down button arrangement affords an operator the ability to select CD tracks. For most music programming, CD players are set to play only one track at a time, but selector switches may permit continuous playback of multiple CD tracks if preferred.

Some professional CD players utilize trays to load compact disks, similar to those found in consumer players, but a number of other models accept plastic cartridges containing individual CDs. When a CD cartridge is inserted into a CD player, a small, spring-loaded sliding door is pushed open, exposing a portion of the lower surface of the compact disk to the player's laser-reader. Cartridges are used to protect CDs used in the hectic environment of radio control rooms. The cartridge system reduces the likelihood of a CD's playback surface becoming contaminated by fingerprints and coffee spills or being damaged in other ways. A compact disk that has been scratched or otherwise fouled can lock up in the middle of a track or skip from the middle of one track to another during playback.

Digital recorders such as the 360 Systems Shortcut have found a place in many control rooms and broadcast studios as a replacement for older, analog, reel-to-reel tape-recorder decks. Units like these permit the operator to record telephone conversations or field reports from a station's news, weather, or sports reporters, edit them as needed, and play them on schedule. As another example, a host of a music program might record a telephone caller's song request and then locate the recording while the program continues. When a tune is ready to be played, the host can play back the caller's request followed immediately by the recording itself, resulting in a nicely produced effect.

Transport controls on these devices tend to be familiar, with the conventional play, pause, stop, fast-forward, and rewind buttons. Other controls reflect the recorder's uniquely digital features, for example sampling rates and waveform editing. The unit might additionally include features such as gain adjustments for input, output, and monitor, as well as VU meters, menu-selectable functions specific to the recorder, and an integrated keyboard for typing file labels or entering commands.

In the 1990s, computer-based **digital cartridge systems** began appearing in radio stations. Even though digital cart systems bear the name "cartridge," the only similarity between them and their analog-tape counterparts is in their function within the radio station. These digital carts have supplanted continuous-loop analog tape cartridges that had been a standard radio tool since the 1960s. Now as then, carts are used for playback of commercials, public service announcements, jingles, news actualities, and often entire musical pieces dubbed from phonograph records and CDs. Although the same kinds of audio materials are recorded on digital carts, they are stored on a computer hard disk instead of tape. On-screen icons in these systems serve the same purpose as paper labels on analog carts.

For maximum hands-on control, operators may elect to start each cart manually, either from the mixing console or from the computer. Otherwise, the operator may place the system into a fully automated mode, allowing it to play each audio file without operator intervention in a preprogrammed sequence. In the middle of the continuum from fully automated to fully manual operation is an approach known as **live assist**, which involves a blend of automation and operator intervention. Live assist permits the host of a program to play a series of items such as songs, jingles, or brief promotional announcements (drops) in automated mode, halting the system for live announcements. Upon concluding those announcements, the operator will return the system to automated mode for the next set of items. This sequence can be repeated as long as needed, shifting back and forth between modes of play as needed.

Even though digital audio equipment has replaced much of the analog equipment found in radio stations, analog devices such as turntables, reel-to-reel tape decks, and tape-cartridge machines have not completely disappeared from the scene. Much material in formats they can play is still usable for broadcasts, and so they remain useful as production tools.

Depending on a station's programming, its control room might be equipped with other facilities, such as satellite receivers for network program material, two-way communications radio receivers for monitoring police and emergency radio transmissions, or a telephone system integrated into the console used for telephone interviews or field news reports. Still other equipment that may be found in the radio broadcast studio include the following: a VHF or UHF radio receiver for reception of play-by-play audio from cross-town sporting events or other remote broadcasts; a computer with Internet access for retrieval of MP3 audio files such as weather forecasts voiced by out-of-town meteorologists; or a television receiver for rebroadcast of TV audio under standing arrangements with a television station or cable channel. In addition to acting as an audio source, TV receivers can be used to alert the radio staff to fast-breaking developments in news, sports, and weather events. Local weather cable channels are especially useful, given the importance of up-to-the-minute weather information.

One final control-room audio source should be mentioned, since the Federal Communications Commission mandates its provision in U.S. radio stations—the Emergency Alert System (EAS) receiver. The EAS receiver fills the dual roles of audio source and information source. When the Emergency Alert System is activated for national, state, or local emergencies such as severe weather, the station may rebroadcast audio from the EAS receiver either automatically or under manual control by an operator. Alternatively, personnel on duty might simply use the EAS receiver as an information source, allowing local staff to voice the alert details on-air. The Emergency Alert System is also activated in some states for Amber alerts, used to broadcast time-sensitive information related to the search and rescue operations concerning abducted children.

ADDITIONAL IN-STUDIO INFORMATION SOURCES

The air studio/control room serves as the central hub for all sorts of radio station programming activities. To fulfill this function, it is helpful to make a range of information sources available to staff working in the room.

The abandonment of noisy Teletype machines in favor of computer delivery of news wire services has made it possible for these sources to be moved into the studio. In-studio newswire access—such as the Associated Press and Reuters—gives air personnel direct access to news bulletins, sports scores, weather information, and many other items of interest. Information on a wide variety of topics is thus constantly at operators' fingertips. Interestingly, the silent delivery of news copy has led a few radio stations to employ the recorded sound of old Teletype machines played softly during newscasts, in order to create an authentically newslike sound.

Modern radio studios are likely to have available one or more computers for Internet access, in addition to computers used mainly for playback of recorded program items. Many audio resources are available on the Web that may either be directly aired or transcribed for voicing by a local announcer. In-studio Internet access makes it possible for talk show hosts to respond to email from listeners and for music show hosts to respond to song requests received by email.

THE RADIO PRODUCTION STUDIO

In addition to a primary air studio and control room, most radio stations provide one or more production studios, used to prerecord audio materials for later broadcast. Some production studios are designed as near duplicates of the primary control room/air studio and can quickly be converted for use as an on-air studio, in case of technical failures in the main facility. More often however, the production studio tends to be a scaled-down version of the primary control room with an emphasis on its recording functions.

Since modern radio production depends heavily upon multitrack digital audio workstations that can handle complicated mixes and signal processing, the producer normally has little need for a large mixer with many playback channels. In truth, consoles in today's production studios have largely taken a back seat to the DAW, with the workstation becoming the primary center of production activities. Often the production console does little more than route audio to and from the DAW to other recording devices. In the production studio these devices are likely to be tape, compact disk, and the station's digital cart system. In studios having such arrangements, the digital audio workstation is depended on to perform many of the chores formerly assigned to the mixing board and outboard signal processors.

The production studio DAW might take the form of a combined hardware-software system geared primarily toward commercial spot production or of a software-only editor that functions well for both short-form and long-form productions. Among such hardware-software solutions are machines like the user-friendly Orban Audicy, which can accommodate up to twenty-four channels of audio. The Audicy also offers the convenience of an outboard, hands-on control surface having real faders, not just the on-screen faders typical of most software-only editors. The Audicy shifts much of the processing function from the computer's CPU to its own processing cards. Add-in cards also provide the RAM necessary for the operation of the Audicy and currently provide enough recording capacity for complex radio spot productions using all twenty-four tracks.

Another company, Digidesign, has long offered combination hardware-and-software digital audio workstations based on its Pro Tools software. This firm continues to offer a number of different production packages useful to both radio stations and recording studios. In recent years, the Digidesign Digi-002 has found its way into radio station production settings by offering, among other things, an external mixer control surface with automated faders.

One of the Windows-based software-only editors that frequently can be found in radio stations is Adobe's Audition, formerly known as Cool Edit Pro. Audition is particularly helpful in the radio production studio, not only for creating multitrack productions and processing, but also for conversion of audio to MP3 files for Internet distribution. Advertisers often contract with a lead station for production of their commercials, and when the spots have been readied they must be distributed to other stations that will also broadcast them.

RADIO NEWS FACILITIES

Radio stations that present a substantial amount of news programming must maintain newsrooms for newsgathering and preparation. These are usually separate from news booths, which are tiny studios used for news production. Still today, radio news anchors usually broadcast from soundproofed, acoustically treated news booths or studios, but staff reporters might voice live reports from the newsroom. This is done to convey an impression of a busy newsroom, complete with the sounds of ringing telephones, police scanners, and the conversations of other news staff.

In a modern radio newsroom, as in so many other areas of a station, computer workstations are essential. Computers are used in the newsroom not only to write news stories but also to produce news features. A typical newsroom includes multiple computers for accessing news wire services, accessing the Internet, and writing and editing news copy, as well as recording and editing audio pieces. A reporter's workstation might include a small audio console, some type of digital audio editor, and inputs from a selection of audio sources such as microphones, a radio news network, and other recording and playback devices. One audio source that finds heavy use in all newsrooms is the telephone. In addition to their use by reporters in conducting investigative work, telephones allow interviews to be recorded on workstations from which sound bites can be extracted.

News booths are usually equipped with modest audio consoles connected to devices such as cartridge machines, network feeds, two-way radios, digital cartridge systems, integrated news authoring systems, and so on. Interconnections among different reporters' consoles permit a news anchor to engage in live discussions with reporters working on breaking news. Since radio networks such as ABC offer multiple feeds simultaneously—one or more channels for regular network newscasts or news feeds for affiliated stations, and other channels for long-form programming such as extended event coverage—news booths should have access either directly or indirectly to all available channels. And, as in any studio, news booths and newsroom workstations must offer suitable monitoring capability via loudspeakers and headphones.

It should be evident that any contemporary radio news operation relies heavily upon computers both for accessing information and for audio production of news broadcast material. Not only do computers provide access to a wide range of services, but they also enable a station to share news stories and associated audio clips with stations in other cities. This is especially useful in exchanges between stations in different locations owned by the same company. The flexibility afforded by these linkups is so great that by using appropriate software it is possible for news staff in one city to produce complete newscasts for broadcast in distant cities. News personnel can simply record, or digitize, a newscast using their local workstation system, saving audio to a file given a name associated with the second station. An automated system operator of the out-of-town station can then download the prepackaged newscast for broadcast at its scheduled time. In an arrangement like this, fewer news personnel are needed, and this translates into reduced operating costs. Under such a centralized system, a news staff of twelve to fifteen people may be sufficient to handle the bulk of news responsibilities for several cities. Only a skeleton news staff at each out-of-town station would be needed to gather local stories and related audio on behalf of the central newsroom.

Despite the current emphasis on digital technology in radio, analog tape still enjoys a place in many newsrooms, in the form of standard audiocassettes, broadcast tape cartridges, and reel-to-reel machines. The news reporter may record finished news segments onto analog tape cartridges for playback within a newscast or embed the audio directly in computerized news scripts, if an integrated word processor, newswire, and audio editing software package such as NewsReady 32 is used. The latter approach allows the news anchor both to read the news story and to play audio directly from the computer in the news booth, eliminating the need for paper copy and audio tapes.

BASIC RADIO PRODUCTION CONSIDERATIONS

Computer technology has dramatically changed techniques of radio production in the production studio, in the newsroom, and in the on-air control room. Especially at music-intensive stations that have adopted digital cart systems such as Prophet Systems' Innovations NexGen or Broadcast Electronics' Audio Vault, on-air duties for control room operators have eased considerably. Even so, aesthetics of radio production have not changed as much as the tools employed to achieve them. To a large extent, the nature of a radio station's on-air sound—both in terms of audio quality and apparent effortlessness of execution—is predetermined by operational and programming policy. No matter the nature of the station's on-air programming, it should sound well produced. The following brief discussion considers two of the many factors that lend programs a pleasing professional impression.

For a start, levels must be appropriate for broadcasting purposes. As detailed above, this means that regular and recurring audio peaks should fall between approximately 85 and 100 percent on the console's VU meter. It is worth emphasizing again that while standard engineering practice calls for the use of audio compressors or limiters to keep audio levels within bounds, good practice dictates that levels set by the control room board operator be as consistent as possible.

Another factor that contributes to overall production quality is the manner in which the assorted audio elements are assembled. For all radio formats, "dead air" is simply not permissible. Dead air is an unintended silence or gap that occurs between production elements. Dead air imparts to productions a loose, poorly prepared, and amateurish quality. For most formats, one audio element should segue, or transition, smoothly and immediately to the next. And for popular music formats, an audio element should begin before the preceding element has completely faded to silence, thus creating a momentary overlap of sounds. This technique affords the program a rapid pace. In more energetic music formats such as contemporary hit radio and oldies formats, the operator would normally start an audio element somewhat earlier, even before the first element has faded much. A board operator for a hit-oriented music format might begin a second song when the first song is just beginning to end, and at the same time manually fade the first tune out to avoid a prolonged overlap of the two songs. This technique raises the speed of a program's flow and maintains its intensity level.

DIGITAL CART SYSTEM BASICS

Generally speaking, modern digital cart systems are more than simple audio production tools. They usually incorporate scheduling modules that permit the station's program director, music director, and traffic director to create broadcast schedules. These schedules provide for music, jingles, commercials, and other elements of programs, and together they are merged into the system's master program log. Once all audio elements have been recorded and saved with unique file names, the system automatically plays the program elements sequentially as specified by the master log. As a program proceeds, the system creates a record of each item aired, simplifying the task of authentication for billing of advertising clients for commercials that have been broadcast.

For obvious reasons, most production duties involving digital cart systems take place in the production studio well before actual transmission. In a totally automated station, production staff must digitally record all audio elements, including each and every announcement, saving them on the system's hard drive. Stations that use the system in a manual or semiautomated mode—that is, live assist—can selectively record on the system only those needed elements, allowing other items to be played from other sources such as CD players, tape cartridges, and the like. In a pure manual arrangement, the board operator individually starts each commercial, song, jingle, or other program element. Selector buttons on the mixing console allow the board operator to play individual audio files from the computer system as desired.

In a hybrid arrangement, an operator arranges a number of songs, jingles, or recorded positioning and promotional announcements to play in automated mode, with the system pausing at a stop cue placed within the master log. At the indicated point, the announcer may insert live talk. When the announcer finishes, the operator can restart the system, perhaps playing a series of commercials; a **stop set**; or another series of songs, jingles, and brief positioning elements (commonly referred to as a music sweep) in auto mode.

Many stations rely on satellite networks for some portion of their programming, whether music, sports play-by-play, or talk shows. The typical satellite network embeds inaudible automation cues in its programming that trigger specific events at automated stations. For example, one cue might silence the network feed at an appropriate juncture to initiate playback of local commercials or other materials, while another cue would un-mute the network feed to allow local stations to rejoin network programming. A different cue might trip brief local announcements that make a nationally distributed satellite program sound more local, and yet another cue might trigger playback of the legal station identification (ID) announcement.

A legal station ID consists of a station's FCC-assigned call letters and the city of license. Anything else substituted for these cannot qualify. Failure to transmit these identifications hourly can subject the station to citations and fines from the Federal Communications Commission. As an example, the legal station identification for hypothetical radio station WXXX in Detroit must include "WXXX, Detroit." (*The Code of Federal Regulations*, Title 47, Volume 4, part 73, section 73.120 provides additional information relating to legal station identification announcements and is accessible through the FCC's website: http://www.fcc.gov.) A station's slogan or positioning statement, "Triple-X Radio" or "Triple-X Radio, Detroit" is not considered a legal ID. Furthermore, a least one legal ID must be aired for each hour of broadcast, as near to the top of the hour as programming permits. For the remainder of the hour a station may air as many slogans or positioning statements as it likes.

Even in large markets, many stations operate at least part of their broadcast days—particularly evenings, overnights, and weekends—without an operator present, relying totally on the automation system to do the job. If need be, those who are responsible for creating commercial and music logs can produce in advance several weeks' worth of programming, permitting a system to run unattended for a considerable period of time—though, of course, this assumes that all required audio has been saved and is available in the digital cart system. Some radio stations are completely automated twenty-four hours a day seven days a week, and these stations often have no one present in the control room a large portion of the time. This is a growing trend among stations, even in major markets, as technology improves and station management attempts to shrink employee-related expenses.

In these automated systems, it is doubly important for producers to maintain consistent levels when dubbing audio into the digital cart system. Otherwise, differences in levels between adjacent audio files can make an on-air broadcast sound highly unprofessional, particularly if no operator is on duty to correct playback levels. Automation systems generally include voice tracks, which are recordings made by the program hosts, introducing or back-announcing songs, promoting the station and its events, and so on. When laying down voice tracks, it is important that they be recorded at robust levels; otherwise, an announcement might be overpowered by stronger audio playing at the same time, for example, playing over a song's introduction. Even if the digital cart system has a ducking option to reduce the gain of music or jingles playing under the voice track, ducking cannot be depended on to correct weak audio. Apart from voice cuts, other audio sources required by automation systems include music recordings from the playlist, jingles, commercials, and public service announcements.

Commercials and weather forecasts produced elsewhere regularly arrive at radio stations as email attachments in the form of MP3 files. The MP3 recordings employed in broadcasting seldom have bit rates less than 128 KBps. In fact, 128 KBps seems to be the bit rate most often used at the local level, even though producers sometimes employ higher rates as a hedge against quality losses that occur in the process of transmission. Some producers also prefer to compress audio files using the MP3 Pro algorithm, which claims better high-frequency response than normal MP3 compression.

Regardless of the material entered into a digital cart system, there are several factors that operators need to take into account. When dubbing audio to a digital cart system, operators should take care to trim any silence remaining at the beginning of each file, either by manually editing or using automatic trim settings in the system's editor. Otherwise, the silence would remain a part of the file, taking up unnecessary hard-drive space, delaying the audio's start once triggered, and making for a sloppy production sound.

Another important aspect of dubbing audio into a digital cart system is proper placement of the end-of-message (**EOM**) cue at the audio waveform's termination. The EOM cues the system to trigger the next event. In order to ensure a consistent, tightly produced on-air sound, the EOM should be placed so that the following event starts with a slight overlap at the end of musical recordings, or at least immediately at the end of spoken word events. The aim is to generate transitions that are quick but not abrupt. However, extreme caution should be exercised to avoid transitions that cause playback of speech overlapping other speech, producing the effect of two persons talking at once.

Finally, even in modern radio stations that rely heavily on digital cart systems for their operation, a variety of additional audio sources, described elsewhere in this text, often come into play. Among the pieces of equipment that radio production personnel encounter might be CD players and recorders, digital audio workstations, as well as older analog technologies such as reel-to-reel tape decks, cartridge tape recorders, and phonograph turntables.

CHAPTER FOURTEEN

FIELD RECORDING

When recording outside a studio, that is to say, "in the field," engineers may face many different kinds of production possibilities. **Field recording** might involve catching a monaural sound bite for broadcast news, location sound for film or television, or the multitrack recording of a concert. It might even include unusual projects such as capturing wildlife sounds for research (called bioacoustics) or for sound effects. In short, the rubric of field recording encompasses any audio production not based in a studio, whatever its purpose. The following pages will present some techniques and information on equipment specific to this type of audio production.

PREPRODUCTION PREPARATION

Preparation is the key to success in field recording, regardless of what is being recorded or where the work is being done. When audio producers venture outside the security of a studio, where backup systems, tools, and resources are close at hand, they enter an environment that is no longer predictable. The surest way to ward off problems from failures when working in the field is redundancy. This is especially important for items that are prone to malfunction, such as cables, batteries, recording media, and the like. If the job requires three microphone cables, the field kit should include six; if the job requires three microphones, at least four should be taken to the field. Other suggested provisions are plenty of spare batteries, disks, storage media, and tapes, as well as multiple power-extension cables.

Lists can be very useful in keeping track of required items. Devising setup diagrams and packing lists will help guarantee that every single piece of equipment needed on location will be available. Lists must be double-checked before departing the studio or equipment room. Checklists are even more valuable when a field activity will be repeated. The first time a producer sets up onsite is the time to make sure that the list is accurate and thorough. Once verified, technicians can be confident that everything needed is on the list the next time out.

A second thing to take along on all field-recording exercises is patience. Field recording will almost always take longer than anticipated, especially if the location is new or one that has not been thoroughly scouted. It is best to make arrangements to arrive early and to stay late; allow plenty of time for setup, run-throughs, retakes, unexpected downtime and breakdowns. Running short on time is bound to lead to compromises, frustrations, and disappointing results.

ACOUSTICS IN THE FIELD

Perhaps the single greatest challenge in field-recording situations is managing the acoustic environment. For comparison purposes, consider the recording or broadcast studio: It presents idealized conditions where sounds can be isolated and reverberation controlled. Since those conditions rarely exist outside of the studio, field engineers must accept compromises in order to produce acceptable results. However, special tools and techniques can be employed to minimize the effects of concessions to environmental conditions.

Windscreens are devices that attach to microphones to reduce noises produced by wind and other rapid air movements. Examples can be seen in Figure 14.1. Two popular windscreens are known by the trademarked names Windjammers and Softies. The Windjammer is a rigid enclosure lined with an acoustic foam material into which a microphone is placed. The open-cell foam material used in its construction is highly effective in baffling wind noise. Softies are also constructed of open-cell foam, but are covered with a synthetic "fur" that serves to further baffle wind noise.

Close-miking techniques can help decrease pickup of room resonances and aid in reducing extraneous sounds in a noisy setting. When positioning a micro-

FIGURE 14.1   The rigid enclosure of the Windjammer, on the right, shields the microphone with acoustic foam material. The Softies windscreen, on the left, also employs acoustic foam material inside a fuzzy exterior that is intended to baffle wind noise.

phone close to a sound source, the desired sound source rises in level much more than distant sources, lessening the likelihood that unwanted noises will intrude noticeably. Close-miking is useful in recording **sound bites** at loud news events or live musical performances, and in capturing natural sounds to be used as sound effects.

For purposes of isolating sound sources, hypercardioid or super-cardioid microphones can also be utilized. These types of microphones are especially handy for capturing location sound in film and television productions where close-miking cannot be employed because the microphone would be visible on-camera. The patterns of hypercardioid microphones are restricted to a small pickup angle by the internal addition of a long **interference tube** that focuses acoustic energy at the mic's diaphragm. The narrow pickup pattern isolates sound sources, separating them from other environmental noises. In video and film productions, the directional microphone enclosed in a windscreen is commonly mounted on a glass fiber or aluminum "fish pole" to allow operators to move it near sound sources. The microphone is held just outside viewers' sight, usually just above or just below the edge of the camera frame.

RECORDING MUSIC OUTSIDE THE STUDIO

Musical events are among the most demanding location assignments audio producers are likely to face. Thankfully, a set of techniques has evolved that provide a template for successful music recording outside the studio. The least complicated approach is to take advantage of existing live sound systems already in place or installed by a touring company or sound contractor. Feeds from a house system or from microphone splitters, as described in Chapter 12, Live Sound Reinforcement, can serve as the foundation from which a complete recording can be created. Such feeds can be augmented with stereo microphone pickups or dedicated mics in selected locations, to flesh out the sound image. Of course if no sound system is available, all the equipment for the recording will have to be brought to the site, after which the project can be treated more or less as if it were a live sound recording session.

Frequently, live performance recordings are made from a series of musical performances that are merged during the mix-down to a stereo master. When this is done, it is necessary to ensure that each piece is performed at exactly the same tempo so that tracks can be properly matched. To do this, **click tracks** can be routed at each performance to one or more of the rhythm section's in-ear monitors. In the mix-down, the best performance of each musician can then be compiled, just as in a studio recording.

At the advanced end of the complexity continuum is the use of location-recording trucks. These **remote trucks** are really control rooms on wheels, bringing many of the benefits of the recording studio to the performance site. Remote trucks offer many advantages. They come in all shapes and sizes, running the gamut from small vans with a modest console and a recorder to forty-eight-foot tractor-trailer rigs with video monitoring, automated digital consoles, and the latest multitrack recording technology.

The chief reason audio producers might prefer to use an on-truck system is that they offer an extensive selection of standard equipment, already wired and tested. That fact gives producers peace of mind and greatly shrinks setup time. It also means the system can be counted on to provide dependable results. Furthermore, these trucks are staffed by knowledgeable technicians, people who are able to troubleshoot problems quickly. Finally, a remote truck parked outside the performance venue will be acoustically isolated from the performance. This and the use of studio-quality monitors enable the truck to offer a familiar and consistent listening environment, reducing the possibility of unpleasant surprises at mixdown time when tracks are combined in another studio.

In the usual case, each instrument and performer will be miked, and the microphone attached to a **splitter** box, as detailed in Chapter 12. Splitters are sometimes included in the stage end of snake boxes. Other splitters allow multiple snakes to be connected via multi-pin connectors. The splitter divides and isolates the signal from each microphone through the use of transformers. Individual signals are sent to the venue's house mixing console (called front of house, or FOH), another is sent to the truck, and sometimes a third is sent to a console for producing a monitor mix.

Several manufacturers offer **digital snakes**. In the digital snake, audio signals are converted from analog to digital at the stage box, then sent to the console or consoles over Category 5 twisted-pair network cabling, or possibly optical fiber. Digital snakes provide tremendous flexibility in layout and signal routing. For instance, virtual channels can be assigned as mic channels, unbalanced or balanced line level channels, or returns. The signal similarly can be split as many times as the situation requires, providing separate signals to multiple consoles and recording devices. Most digital snakes supply a computer interface through which a user can set gain or attenuation levels on each channel, adjust phasing on each channel, and define routing layouts. A big advantage of a digital snake is the reduced expense of long Category 5 cable runs as compared with costly multiple-pair shielded analog cable. However, this is offset somewhat by the higher cost of good-quality analog-to-digital converters contained in the snake boxes, and the potential for latency problems caused by conversions.

Inside the remote truck, microphone outputs are sent to a mixing console where their signals are equalized and routed onward to multitrack recording equipment. The goal of the location-recording engineer is exactly the same as the studio engineer's: capturing each instrument or voice at an optimum level for later mix-down. Recordings can be made on tape machines or DAWs, depending on mixing and post-production plans. A general layout for a remote truck is presented in Figure 14.2.

Enthusiasm for the use of remote trucks seems to have peaked in the 1980s, when many recording artists embraced the idea of having the studio come to them. But later, as computer-based recording technology developed and became affordable, many of those same artists built home studios and no longer needed re-

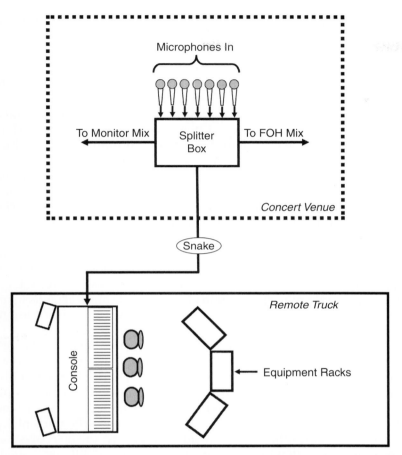

FIGURE 14.2 When a remote truck is used for live recording, a snake connects concert venue microphones with the console located outside the performance site.

mote trucks. A parallel shift has occurred in the creation of live concert recordings. Rather than using remote trucks, they now tend to be made from pre-fader signals sent directly from the FOH console to a multitrack digital workstation setup in an acoustically isolated spot within the performance area. Remote trucks nonetheless remain an important option for production of major artists' live concert recordings and for audio support of broadcast events.

STEREO RECORDING METHODS

Symphonic music production requires a different approach to mic placement than that employed for contemporary popular music such as rock, country, or blues. Because a full symphony orchestra may consist of as many as ninety instruments or more, and because a concert hall becomes, in effect, a part of the orchestra's sound, close miking of individual instruments is generally not practical or desirable. Instead, whole orchestra sections may be miked, and recorded separately on multitrack systems, and those tracks can be mixed down to a master recording later. At the mix session, the engineer can properly set levels of each string section, blending the strings with the brass, woodwind, and percussion sections.

Purists sometimes contend that the only way to capture a natural orchestral sound is through use of single-point stereo techniques. This approach makes use of a stereo microphone, pair of microphones, or **binaural dummy head** microphone placed in an optimum position within a performance hall. The stereo mic output is normally recorded as cleanly as possible without use of effects.

Stereo microphones are available containing two capsules or elements within a single case, as briefly explained in Chapter 4, Microphones. The two elements must be carefully matched and properly positioned together to capture discrete left and right images. A few companies manufacture true **binaural microphones**, most often in the form of dummy head mics. As the name suggests, these are manikin heads that have a pair of small microphones placed in their pseudo ear canals. The dummy head method delivers a recording that may offer the most authentic reproduction of a large-scale musical performance. When heard through headphones, dummy head recordings can deliver a listening experience of astonishing realism.

Pairs of well-matched conventional microphones can be used to produce excellent stereo imaging also. The following discussion will present several ways of making good-quality single-point stereo recordings. The **X-Y coincident technique**—sometimes termed the intensity stereo method—employs a pair of cardioid microphones mounted so that their diaphragms are near to each other, with their lobes separated by an angle of at least 90 degrees. The microphones can be placed one above the other or side by side. Either way, the objective is to position the microphones so that their diaphragms pick up two different sound fields, as shown in Figure 14.3. Some professionals use the term X-Y to refer only to the 90-degree arrangement, and refer to the 135-degree method as *coincident.* Most single-unit stereo microphones have their two capsules positioned in a coincident arrangement within their cases. Coincident mics are said to offer superior stereo separation, but they do not spread the signal completely across the stereo field; the signal lacks spaciousness.

In the 1930s Alan Blumlein refined the coincident technique by using two bidirectional microphones with their diaphragms positioned 90 degrees apart so the lobe of one fell into the null of the other. Audiophiles often credit the **Blumlein technique** with sonically accurate recordings, ones that contain precisely defined stereo images (Figure 14.4).

Another version of the coincident method is the middle side, **mid-side**, or **M-S**, arrangement. A true M-S arrangement includes an omnidirectional or cardioid microphone (the middle), combined with a bidirectional microphone (the side), with the two lobes of the bidirectional microphone connected in a phase relationship so as to cause summing and differencing of the two mics. This is shown

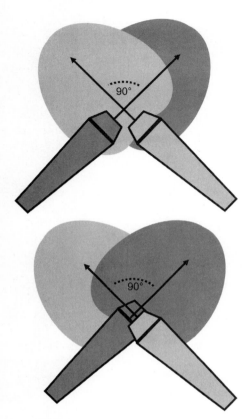

FIGURE 14.3   X-Y coincident stereo mic setups.

FIGURE 14.4   Blumlein stereo mic setup.

in Figure 14.5. To achieve differencing, one lobe of the bidirectional microphone's output signal has its polarity shifted 180 degree as compared to the other lobe. Some microphone manufacturers (e.g., Schoeps) market M-S systems having a breakout box for easy connection of mics in the desired phase relationship. Although the phase relationship is theoretically important, in practical situations engineers sometimes set up M-S stereo mics without concern for their patterns, summing or differencing.

Yet another adaptation of coincident placement is the **near coincident technique**. In this approach, two cardioid microphones are placed with their diaphragms angled apart rather than crossing one another. The Office de Radiodiffusion Television Française, the French national broadcasting organization, adopted a version of the near coincident technique, and their method has come to be known as the **ORTF system**. In it, two microphones are aimed 110 degrees apart and spaced about 17 cm (seven inches) apart. The illustration in Figure 14.6 indicates how this is done. The result, according to many listeners, is stereo that

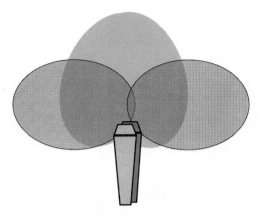

FIGURE 14.5   Mid-side stereo mic setup.

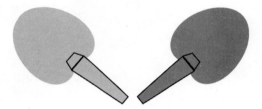

FIGURE 14.6   Near coincident ORTF stereo mic setup.

has outstanding localization, or well-defined locations of sound sources within the stereo field.

The **spaced pair**, or A-B technique, uses at least two microphones spaced as much as ten to twelve feet apart, both pointing toward the sound source. Spaced pair microphones may be of any polar pattern, but omnidirectional mics are employed most often. The spaced pair technique shown in Figure 14.7 produces a stereo image that is often described as spacious, containing a good deal of ambience. This perception is probably induced by phase inconsistencies between the two microphones. Reverberant early reflections will arrive at the two spaced diaphragms at different times, creating phase cancellations and reinforcements that are not present when using coincident mics. Because of the closeness of the two mics used in coincident setups, their phases tend to be similar. Sometimes the spaced pair technique can be improved by the addition of a third, center microphone facing forward, filling what is described as a "hole" in the stereo field.

In the mid-1950s Decca Records pioneered spaced pair microphone techniques, coupled with a third center microphone. This method became known as the **Decca Tree**. Roy Wallace, one of the Decca engineers who developed the system, used a metal stand to mount the microphones. His partner, Arthur Haddy, said the contrivance looked like a Christmas tree, and the name stuck. When used

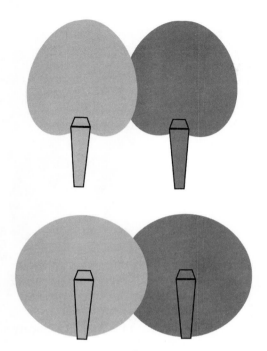

FIGURE 14.7   Spaced pair A-B stereo mic setup.

in symphonic music recordings, the Decca Tree normally employs three omnidirectional microphones placed just behind and above the conductor's position,
about six-and-one-half feet above floor height.

Microphones are arranged on the tree as indicated in Figure 14.8, with the two
outside microphones spaced at least three feet apart. The actual spacing varies
according to the stage width. Some situations may require that outside microphones be pointed slightly outward. The center microphone should be placed between two to four feet forward of the spaced pair. Exact direction and spacing
should be determined by listening to the results. Recording the signals from the
Decca Tree is simple, with the left microphone placed left, the right microphone
placed right, and the center microphone placed in the center for both channels.
Decca Tree recordings present exceptional stereo imaging, and are widely used for
film scoring, orchestral recordings, and for surround recordings.

One final stereo miking technique makes use of two **boundary-effect
microphones**. This type of mic is discussed in Chapter 4, Microphones. The two
boundary mics are mounted on panels measuring two feet by two feet, angled 60
degrees apart (Figure 14.9). The mic elements should be located 6.7 inches apart,
and 6.7 inches from the front edge of the panels. This results in the two mics each
picking up a hemispherical area with some overlap in the center.

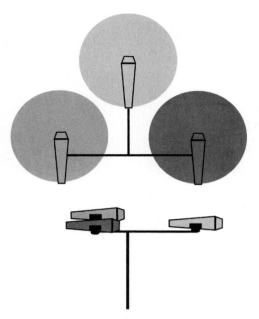

FIGURE 14.8   Decca Tree stereo mic setup.

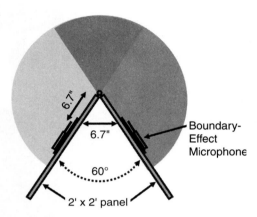

FIGURE 14.9   Use of boundary mics in a stereo setup.

As just noted, these minimalist stereo miking techniques are particularly well suited to orchestral recordings, but their use need not be limited to those applications. The same methods have been applied successfully in recording jazz, blues, rock, and chamber ensembles. All the stereo miking techniques mentioned here have also been widely used in recording sound effects, film sound, and wildlife sounds.

MIC TECHNIQUES FOR THE FIELD

The recording of wildlife sounds, or bioacoustics, is an interesting subfield that often makes use of unusual equipment and novel techniques. Because recording wildlife projects can be tedious, it is also a specialty that calls for producers with staying power. A key tool in this kind of work is the **parabolic microphone**—one that can boost sound levels of distant or soft sounds, while rejecting surrounding ambient noises. As mentioned in Chapter 4, Microphones, these mics are made of a parabolic-shaped dish arranged so that sounds bounce from its hard surface to a focal point where a cardioid microphone is fixed in position. These highly directional mics are favored in broadcasts of sporting events and in electronic surveillance. The drawing in Figure 14.10 illustrates the operating principle of a parabolic microphone. Because of the materials used in its construction, this

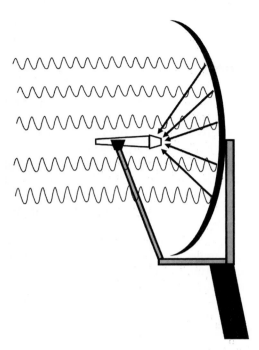

FIGURE 14.10   The disk of a parabolic mic setup focuses distant sound at the diaphragm of a pickup microphone.

kind of mic sacrifices low frequency response in order to obtain directionality and noise reduction.

Another subfield is recording live news events for broadcast purposes. The collection of sound materials for use in news reports can take many different directions. Who hasn't seen a big handheld microphone with its conspicuous call-letter flag in front of an election candidate, or the forest of fish poles with Softies bobbing and weaving while some celebrity emerges from a limousine, or the hodgepodge of microphones that obscures a podium where public officials speak?

For handheld uses in news interviews, broadcast journalists generally choose an omnidirectional dynamic microphone. The omnidirectional mic is forgiving of errors in off-axis placement; its broad pickup pattern can catch a newsmaker's sound bite even when the journalist's aim is poor. Mics designed for this purpose, such as the ubiquitous Electrovoice 645A, are less susceptible to handling noises. For radio purposes, the microphone can be connected with any kind of portable recording device, the most popular types being DAT, minidisk, and solid-state recorders. For television purposes, handheld microphones can be outfitted with an after-market wireless transmitter. Several manufacturers produce compact transmitter-receiver combinations that mount directly on professional camcorders. For TV standup reports, wireless setups can be paired with a lavaliere microphone clipped to the reporter's collar.

Fish poles are a standard part of every news producer's audio toolkit. These microphone mounts do, in fact, resemble a fishing pole, but are constructed from aluminum, carbon fiber, or glass fiber. Their length is adjustable from as little as two or three feet to as much as twelve feet. The microphone is attached to the pole by a **shock mount** that acoustically isolates it from handling noises conveyed through the pole. Most commonly used on fish poles are super-cardioid or *hyper-cardioid microphones*, because of their ability to isolate sound sources. When these mics are used, they must be carefully aimed and operators must be attentive to movements of their target. Microphones like this will naturally be exposed to rapid air movements and thus need windscreens—either the rigid Windjammer type, or the fuzzy Softie. As in all news applications, microphones may be directly cabled to the recording device or coupled to a wireless system.

Large, well-orchestrated national or international news events such as press conferences or political campaign events, are usually easy to capture in sound. Event planners commonly provide a press box or **mult-box**. These boxes are multiple splitters that connect microphones to dozens of transformer-isolated outputs. The output connectors may be XLR, quarter-inch, and RCA types. Broadcasters need only attach their recorders to the splitter to receive audio feeds.

LOCATION RECORDING FOR THE SCREEN

Recording field sound for film or video can be a complex undertaking, often requiring a team of skilled audio specialists who may employ just about every trick in a sound producer's book. This kind of work can be physically demanding too; recording tasks may necessitate standing, kneeling, squatting, or sitting for hours in awkward positions while holding a fish pole, listening through headphones, and simultaneously concentrating on VU meters and moving performers. Chapter 16, Sound for Pictures, discusses three sorts of recorded sound used in film and television: dialogue, sound effects, and music. The discussion here will be limited to capturing sound in the field for these uses.

The standard rig for location sound begins with appropriate cases, bags, and carts that make transporting equipment safe and easy. Portability requires that equipment in the kit be compact and rugged. Gear must be able to withstand transport over difficult terrain and long hours of operation in hostile environments. Figure 14.11 shows a typical field production setting.

Location dialogue is normally recorded using a boom-mounted hypercardioid **shotgun microphone**. The overhead studio **giraffe boom** has largely given way to the lightweight fish-pole mount just described. Wireless miniature microphones may also be used on location shoots. In film production, audio is recorded separately on tape so that mixing and sweetening can be done before eventual dubbing to film. The favorite recorders for film sound have traditionally been reel-to-reel tape recorders such as the compact **Nagra** shown in Figure 14.12. This machine includes important features such as time code synchronization, exceedingly stable tape transports, excellent audio specifications, and rugged

FIGURE 14.11   In a typical field production setting, a directional mic on a boom and a portable recorder are used by a technician to obtain good-quality audio recordings.

FIGURE 14.12   The Nagra IV-S portable reel-to-reel recorder has been a traditional favorite for film sound recordings.

construction. As in other areas of audio production, digital devices incorporating all the same features but having the advantage of superior sound quality and ease of transfer to digital workstations are gradually displacing reel-to-reel equipment.

CAPTURING SOUND EFFECTS

Even though digital technologies have drastically altered the way sound effects are created and modified, most that are heard in films, television, and video games still originate as real-world sounds. The majority of location sounds heard in major motion pictures are added to sound tracks in the postproduction phase, even though they originate in the field.

A few specialists have built successful careers capturing, editing, and marketing sound effects. These days, sound effect specialists rely on digital recording equipment to capture sounds of everyday life, which they catalogue and archive for use in many different types of projects. These sounds range from the mundane—doors closing, cars starting, or dogs barking—to the odd—wild animals, crashes, and explosions. Nothing more is required for this specialized kind of recording than a quality recorder, a first-rate microphone, and a good set of ears.

Maintaining a comprehensive set of field notes will make a field recordist's life much less trying. Following a long field session there may be hours of recordings to sort through back in the studio, and relying on memory alone will surely lead to errors. Each **take** should be logged according to its description, start time or position on the tape, approximate running time, and comments. Here is where an assistant's second pair of hands can be invaluable; if things happen quickly, it will be difficult and distracting for the engineer to continue note taking. It is good practice to have field notes transcribed and stored for later reference, on the grounds that one never knows when recordings may be needed again.

CHAPTER FIFTEEN

MULTITRACK RECORDING

Multitrack recording places audio signals on multiple synchronized tracks made either on audio tape or on virtual tracks within a digital audio workstation. The tracks may be recorded simultaneously, or they may be added during separate passes made at different times. Initial sets of tracks are recorded during a **basic tracking** session. A complete recording is obtained by adding the remaining audio tracks in later overdubbing sessions. Eventually the engineer will **mix down** all the tracks usually to a stereo signal, but possibly to a mono or surround sound master. Multitrack recording techniques can be used to create many types of finished products such as popular music recordings, motion picture sound tracks, and radio and television commercials.

THE WORK OF MULTITRACK RECORDING

In multitrack recording, a strict division of labor between engineers and producers can often be found. In this kind of work, an engineer's job will usually be purely a technical one, defined as helping artists and producers realize their musical concepts. Generally in multitrack environments, the duties of mixing console operations are delegated to the engineer while the **producer** makes creative and aesthetic decisions. Highly successful producers such as Quincy Jones, Don Was, and Rick Rubin have become familiar household names, but even the most successful engineers are rarely known outside the recording industry.

Perhaps more than in any other kind of audio work, competent multitrack recording engineers and producers must possess refined listening skills. Ear training can be used to develop such listening skills. In ear training, listeners learn to recognize the frequencies of sounds and to recognize the effects of boosting or cutting selected frequencies. Ear training also entails an ability to identify the harmonic content of musical instruments and to judge equalization applied to the sounds of those instruments. Those who possess good listening skills have hearing that is sensitive to subtle variations in the frequency response characteristics of audio setups. Some especially skillful audio engineers are able to distinguish between dynamic cardioid microphones from different manufacturers, based on their distinctive sound qualities. These abilities are important because an intimate knowledge of sounds produced by each mic in the microphone closet is necessary in order to select the best one for any recording task.

It is not only listening skills that matter, however, but also an ability to apply suitable gain, balance, equalization, and effects adjustments to achieve project goals. In this regard, learning the functions of buttons, knobs, and faders is only half of the battle; knowing when and how to use them creatively is the other, more challenging half. Steep drops in the cost of DAWs and production software have brought multitrack recording capabilities within the budgets of nearly everyone, but just having access to the tools is no guarantee of success. Good recordings still depend on expert creative judgment.

Good interpersonal skills are required for success in multitrack recording work, too. A music session is likely to involve a number of musicians and a producer, all relying on the engineer to translate their artistic notions into a recorded reality. Creative ideas about music are often difficult to articulate, and concepts in the artist's mind may not be matched by sounds played back through the monitor speakers. That can lead to tensions, made worse by the steadily ticking clock, each tick of which sends the recording session's cost upward. Commercial sessions may include the contributions of writers, producers, clients, agency representatives, and assistants, each of whom will seek to lend their creative vision to the project. It frequently happens that an engineer may need to tactfully propose alternatives to the preconceived ideas of a creative team. Even a simple voice-over session may require that the engineer manage sensitive egos gently to encourage the best performance possible from the talent.

AN HISTORICAL PERSPECTIVE

The techniques and technology of analog sound recording were already well established by the 1930s. But at that time, the industry standard was monaural **direct-to-disk** recording. In making monaural direct-to-disk recordings, sound engineers would blend or **mix** sets of microphone signals into a single signal or monophonic channel. This was done in real time as a live performance took place. The process was known as direct-to-disk because the audio signal was routed straight to a **record cutting lathe** for production of a 78 rpm **master disk**. The freshly cut disk could then be used to press copies for distribution to consumers. Figure 15.1 shows how this was done.

The introduction of the audiotape in the mid-1940s offered recording engineers at least two distinct benefits: an ability to erase and rerecord the tape (an economical advantage), and the capability of **editing** the tape (a creative advantage). Tapes could be cut and spliced to remove false starts and—with sufficient practice and skill—to remove or replace whole passages. Engineers could thus change

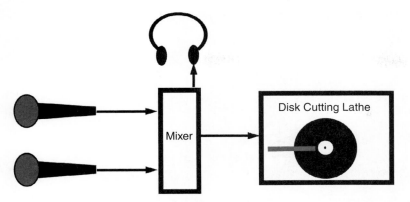

FIGURE 15.1 Early direct-to-disk recording systems placed few devices between the microphone and the disk.

a musical piece's structure and length. Because of its technological advantages, the use of tape recording methods quickly replaced direct-to-disk recording.

Mono tape recordings were made on a single tape track, using the same live process used in direct-to-disk. Engineers mixed microphone outputs into a single signal that was routed to a one-track (monaural) tape machine. This might be recorded on a machine like the one shown in Figure 15.2. A master disk to be used in pressing records was cut from these tapes.

Another big change in the sound recording process came with the introduction of stereo recordings. Stereo or two-track recordings gave more creative options to artists and engineers. In addition to balancing **levels** of instruments and voices, the panning of signals into left and right channels created a stereo **sound field**. Although some experimentation with unusual placements occurred, such as all vocals on the left and all instruments on the right, a generally accepted practice arose in which a sound image of a band or orchestra was created through mixing and panning. This general practice placed vocals in the center front; bass and drums in the center rear; rhythm instruments such as guitars and keyboards either left or right and back somewhat in the mix; and lead instruments up front, panned slightly off center.

In about 1955, Ampex Corporation introduced a three-track tape machine that changed forever the field of audio recording. This was believed to be the first true multitrack tape recorder. The machine was intended to record dialogue, music, and sound effects in motion pictures, but because it offered three independent tracks, it soon found a home in recording studios. Using these recorders, instruments or vocals could be **assigned** to record a single track on the multitrack machine and then later mixed with other tracks to create a stereo **master** tape. Engineers no longer had to make real-time mixing decisions, but could unhurriedly obtain just the right blend of tracks whenever convenient.

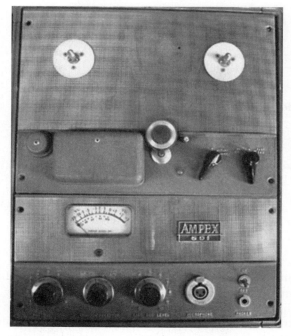

FIGURE 15.2   An example of an Ampex one-track tape machine.

Another key innovation of the three-track machine was overdubbing. In overdubbing, audio material is added to empty tracks following an initial recording session. This permits, for example, a vocalist to overdub lead vocals in a studio long after all instrumental tracks had been laid down.

A curious sidelight of the growing number of tracks was that sound engineers initially installed an extra monitor speaker for each additional tape track. When single-track monaural recordings were standard, control rooms naturally needed only one speaker. As two-track stereo came into use, a stereo sound field was created by the addition of a second speaker. Then, three-track machines led to installation of a third control room speaker. Thankfully, the practice of mounting a control room speaker for each recording track halted well before the arrival of thirty-two–track machines. Eventually, the three-track tape recorder faded into history and stereo mastering and monitoring became an industry standard.

Development of analog multitrack tape recorders accelerated through the 1960s and 1970s. Track configurations grew to four, then eight, sixteen, and finally twenty-four tracks. Tapes developed for use on these multitrack machines grew in width from quarter-inch, to half-inch, to one inch, and finally to two inches. Speeds on some sophisticated twenty-four–track machines were as high as thirty inches per second.

FIGURE 15.3   The Ampex monophonic mixer combined six channels into one.

Concurrent with advances in tape recorders, comparable improvements oc-curred in mixing consoles. Early consoles, such as the one pictured in Figure 15.3, were configured to mix signals from five, six, eight, or ten sources into a single monaural channel. Stereo naturally required that consoles have two output chan-nels, and the shortlived three-track studios demanded three channels. As track numbers grew, so did the input and output capabilities of consoles. The number of output buses expanded to four, eight, sixteen, and ultimately to twenty-four, each one supplying a separate signal to a track on a multitrack tape machine. Large contemporary mixing consoles, like the one pictured in Figure 15.4, may have over one hundred inputs, and as many as sixty-four output buses.

Digital tape technology began to be widely adopted in the 1980s. As noted elsewhere, digital tape technology eliminated tape hiss and other artifacts of ana-log tape recording such as wow and flutter. Digital recorders additionally offered a small, less apparent advantage by placing more tracks on narrower, less-expensive tapes. Some digital machines used half-inch tape for thirty-two–track recordings. By the 1990s, digital tape machines moved to **cassette**, based on var-ied track configurations and cassette housings. Figure 15.5 shows a two-track DAT master recorder (top) and an eight-track R-DAT recorder (bottom).

The next important trend in multitrack recording was the development of computer software-based recording systems, or free-standing digital audio work-stations (DAWs), as discussed in other sections of this book. In computer-based DAW systems, sound is captured through either a **sound card** or an external ADC unit. Whatever hardware used for analog-to-digital conversion, it can be installed in a desktop or laptop computer for audio storage as a **virtual track** on the com-puter's hard drive. Software-based systems offer many improvements over other multitrack types: **nondestructive editing**, economy, portability, outstanding quality, and no **generation loss**.

FIGURE 15.4   Modern complex multitrack mixing consoles provide dozens of inputs and outputs while integrating some signal-processing capability.

FIGURE 15.5   A stereo DAT machine (above) and an eight-track R-DAT recorder (below).

LAYING BASIC TRACKS

Multitrack sessions bear little resemblance to early real-time recording sessions, and over the years they have become somewhat ritualized. It is not unusual for different engineers to work on separate portions of a recording project. A tracking engineer records the basic tracks and may lay down overdubs, and a mixing engineer will perform the final mix-down. In an increasingly specialized music recording industry, some engineers may develop specific genre specialties, and some engineers may even focus on specific recording tasks such as drums or vocals.

Contemporary multitrack recording is often **asynchronous**, or out of sequence. In most instances, **rhythm** tracks, used as basic tracks, are the first to be laid down. For most popular music recording sessions, drums, bass guitar, rhythm guitar, and perhaps keyboards are included in the initial set of tracks.

To keep accurate time, sessions usually begin by recording a click track. A drum machine, sequencer, or other metronome device supplies click sounds. Having the click track available allows musicians to maintain proper time when tracks recorded later are synchronized with previously recorded ones. Any kind of new material can be laid down in perfect synchrony with previously recorded tracks if a click track is available. Moreover, any track, even drum tracks, may be replaced if a click track is present.

During recording of basic tracks or the tracking part of a session, the tracking engineer will attempt to capture each instrument at its maximum undistorted level to keep audio as high as possible above the inherent **noise floor**. In analog recording, maximizing the signal on tape is especially critical, because a strong signal will produce the greatest signal-to-noise ratio. But in maximizing signals, engineers must not **oversaturate** the tape because this will cause audible distortion. In digital recording the noise floor may be less of a problem, but optimizing signal levels will still provide the greatest dynamic range. It is vital not to fall into the trap of laying down tracks at the level they might be in the final mix. Of course, a multitrack tape recorded with each track at its optimum level will sound like a cacophonous mess, with all the instruments equally loud and vocals obscured by instrumental tracks, but mixing will take care of this problem later. To make the recording process easier on the ears, many consoles provide a monitor section that enables creation of a control room mix. This is only a rough mix for control room monitoring purposes, and will have no effect on levels of recorded signals.

Obtaining each track's widest dynamic range begins at the console's first gain stage. On most professional and semiprofessional mixing consoles, a gain potentiometer or **gain trimmer** is included at the input stage of each channel. This control is undoubtedly the most important in determining a recording's dynamic range and signal-to-noise ratio. For a full explanation of this issue, refer to Chapter 5, Mixers. As described in that chapter, the gain potentiometer should be set so that the clip LED is just begins to blink intermittently. Of course, there are minor variations from mixer to mixer in the sensitivity of clip indicators, and so engineers need experience with each particular console to learn how much blinking should be expected before distortion begins.

It is highly advisable to avoid applying any effects such as **reverberation**, delay, or flanging while laying basic tracks. Effects recorded at this point are irreversible. Consequently, tracks should be laid **dry**, and effects added only at the mix-down phase of the recording process. A similar caveat holds for **equalization (EQ)** as well. The guiding rule in basic tracking is that equalization should only be used to ensure that frequencies present in the signal are kept well above the noise level. Later, during mix-down, frequencies can be boosted or cut—but this is only possible if those frequencies are clean and at a good level. It is true that some engineers use a more heavy-handed approach in equalizing drum tracks. In much of contemporary rock, country, pop, and hip-hop music, drums dominate recordings. Achieving the currently fashionable "hot" drum sound necessitates a boost well beyond the norm in certain frequency ranges.

Since tracks can easily be replaced, temporary **scratch vocal** tracks are commonly included when laying basic tracks. This is helpful for musicians in keeping them aware of where they are in the musical piece and in responding musically to vocal and solo performances. These and other critical passages—for instance, horn or guitar solos—can be erased later during the overdubbing phase.

Most recording studio consoles provide a means for studio musicians to hear themselves as they perform. This is essential for overdubbing because the artist may be alone in the studio performing in accompaniment to previously recorded tracks. The mixer section employed could be the cue, monitor, or **foldback bus**. Occasionally engineers use an **auxiliary send** for performers' monitoring. Each input channel of the console will have a send through which individual signals are routed to a headphone distribution system. Although it is possible to route the signal post-fader, headphone signals are most frequently routed pre-fader.

By employing an arrangement like this, each channel's level send can be individually set so that musicians will hear a **rough mix** of the performance in their headphones. In this rough mix, the lead vocalist's microphone level will likely be highest, while rhythm guitar levels will be lower, and drums and other instruments as well as backup vocals will be at appropriate levels. The headphone mix might crudely approximate the final mix, but the primary objective in setting levels is to give artists the balance they need to gain their best performance. The signal going to the recorder will be the same maximized level for each track, so musicians would not want to monitor themselves by that undifferentiated mix.

Some studios can accommodate multiple cue mixes so that performers can tailor their individual headphones mixes according to personal tastes. This is a big advantage because performers typically want their own instrument or voice higher in the mix than would normally be proper for others in the group.

MICROPHONE PLACEMENT IN MULTITRACK SESSIONS

Microphone techniques for multitrack recording derive from the logic of laying tracks that contain isolated sound sources uncluttered by nearby sounds and devoid of artifacts. Studio microphone technique is as much art as it is science; there are really no hard and fast rules to obtain the best results. Volumes have been written on studio mic techniques, and these can be highly useful, but in the end a producer's own hearing and taste should guide microphone selection and placement. Nevertheless, the authors offer a few points for consideration, drawn from their own experience, at the conclusion of this chapter.

Generally speaking, engineers prefer a **close miking** technique in the studio. The method is simple: a cardioid microphone is placed as near to each instrument as is practical. The notion is that a cardioid microphone will reject sound from adjacent instruments, and, by miking each instrument closely, the microphone's gain on the console will be reduced and the target sound source will rise in level compared to nearby sources. Close miking also minimizes the pickup of room reverberations. For any given room there is a critical distance for microphone placement, which is the point beyond which reflected sounds bouncing from room walls begin to overpower the sound coming directly from the instrument or performer. It should be apparent that treating room surfaces with absorptive materials to reduce reverberation will extend the critical distance. The more reverberant the room, the closer the mic must be to its sound source. Results can be further enhanced by the use of noise gates, as described in Chapter 9, Audio Processors and Processing. A gated microphone will be muted unless the level of its intended sound source activates its gate. This yields a track free from stray pickups of unwanted sources.

When selecting and placing a microphone, engineers should take into account the frequency range and dynamic range of each instrument, where sounds emanate from the instrument, and the position from which a listener normally hears the instrument. Experience teaches producers which microphones can tolerate high sound pressure levels, and which microphones' frequency response characteristics complement selected instruments. For instance, high SPL instruments need a more tolerant dynamic microphone, and bass instruments will benefit from a larger diaphragm microphone. Producers consider the design of the instrument and how the quality of sound coming from it differs depending on where a microphone is placed. The sound of an acoustic guitar taken from a microphone aimed at the hole in the guitar body is very different from sounds produced by a microphone aimed at the base of the neck. Similarly, the sound of a drum kit is different from the player's vantage point than from the listener's. Experimentation and critical listening are the keys to selection of the correct microphone and microphone placement.

Miking a drum kit presents a set of special problems. Here, close miking is especially important because of the tremendously high sound pressure levels created by drums and other percussion instruments in the kit. This practice must be tempered by measures to control phase problems introduced by using several microphones in close proximity to one another. As has been noted previously, a sound reaching two different microphones at slightly different points in time will be slightly out of phase. In the worst case, signals can be at or near 180 degrees out of phase, hence producing cancellation. Some authorities rely on the **3-to-1 rule** described in Chapter 4, Microphones, to reduce the risk of phase cancellation. This rule requires that if a microphone picks up sound from a source intended for another microphone, it must be at least three times farther from the sound source. For example, if a microphone is placed one foot from a vocalist, no other mic should be closer than three feet from the singer.

The fundamental principle in laying tracks is that it should yield the natural sound of instruments in the most accurate way possible. A too-often overlooked factor in achieving satisfactory results is the importance of starting with the right sounds. Drum kits with good-quality shells having properly tuned heads, free of hardware squeaks and rattles, should be demanded by both musicians and producers. Guitars that are properly set up for accurate intonation and free of buzzing strings will naturally produce better recordings. Experienced engineers often must remind musicians that the first step in a recording session is getting the proper sound from their instrument. Effects and equalization can enhance a performance, but should not be used as a remedy for poor instrumentation.

OVERDUB TECHNIQUES

As indicated earlier, overdubbing is the second phase in the multitrack recording process during which lead or background vocals and additional instruments are added or replaced. Portions of tracks may be rerecorded, or whole tracks may be replaced in their entirety. Small performance errors can be punched in during the overdubbing stage, in order to create the mistake-free product that consumers expect from commercial recordings. When well done, no listener will be able to detect that a passage has been replaced.

Executing a seamless punch-in is more art than science on analog recorders. Here is how it works: using a special playback mode that monitors the record head, a musician listens while performing along with the previously taped recording. At a precisely timed moment, the record button is pressed, switching the machine into record mode and causing it to begin recording the new performance over selected tracks. Pressing the play button disengages the record function, and from that point onward the pre-existing performance remains intact. The difficulty in achieving a seamless transition is that each tape machine has its own peculiar timing characteristics, and operators must learn to anticipate switching modes to catch transitions at the exact intended point.

On most digital recorders, including software-based systems, punching-in can be more fully automated. A precise point can be marked in the timeline for the **punch-in** and **punch-out** to occur, and the machine or software will carry out the transition at these points automatically. Most digital systems allow a punch-in to be rehearsed before it is executed. In this mode, the existing track is merely muted while the musician rehearses the new passage.

Another overdub technique used to achieve a perfected performance is known as **comping** (from compiling or composite). Perhaps a vocalist will not enunciate a word clearly, will use the wrong inflection, or will not sing a note pre-cisely on key. In comping, a performance is re-recorded several times on separate tracks. The mixing engineer can then select the best of each performance from among available tracks, thereby assembling a single optimum composite track. In a software-based digital recording system, comping is even easier, requiring only that the best performances be edited onto a comp track.

Technological advances have opened up possibilities for collaborations that would have been unthinkable even a few decades ago. Many studios lease Inte-grated Services Digital Network *(ISDN)* lines from their local telephone company. These lines allow performers in similarly equipped studios practically anywhere to participate in recording sessions in real time. ISDN lines are broadband lines ca-pable of transferring large amounts of data at high speeds. Transmission rates are high enough that there is little latency between signal origination and reception at its destination. Any delays can be easily corrected by the use of digital delay units. By this means, it is possible to have a vocalist in New York add vocal overdubs in a session being engineered in California.

Resourceful engineers developed a way to overcome the limited number tracks available on early analog recorders. This technique is known as **bouncing** or **ping-ponging**. Bouncing tracks is handy on nonprofessional cassette-based home portable studios, and is still useful on digital multitrack tape machines that offer only a few tracks. In bouncing, several tracks are mixed together and the re-sult recorded on an empty track. For instance, working on an eight-track system, an engineer might record three tracks of drums, one of bass guitar, and one of rhythm guitar. Those five tracks can be mixed together and bounced to two empty tracks. This frees up the five original tracks so that they can then be erased and used to overdub other instruments.

There is a serious disadvantage in bouncing tracks—all mixing and equaliza-tion decisions made along the way are irreversible. Once the five tracks are mixed down to two, that mix cannot be altered. One caution should be noted: when using narrow-gauge analog recorders, an empty track should always separate playback tracks from recording tracks; bouncing from adjacent tracks might result in feedback artifacts. Although bouncing is not a satisfactory technique for pro-fessional work, it is perfectly acceptable as a way to expand the capacity of systems used in recording demo and semiprofessional sessions.

The use of track bouncing on digital audio workstations and software-based recording systems allows engineers and producers to tailor mixes to specific audi-ences or media. Tracks may be mixed in different ways and saved to a pair of open tracks as unique submixes or **stems**. This allows jingle producers to create mixes containing vocals, without vocals, and multilingual vocal versions for distribution to different sets of broadcast stations. Music and voice-only stems derived from postproduction television commercial sessions may be used also for radio ad-vertising. In postproduction sessions for film and television, three separate stems (dialog, music, and sound effects) can be mixed and then merged into a final recording.

MIX-DOWN SESSIONS

Once all tracks have been laid down—whether 4 or 124—the next phase will be the creation of an artfully mixed two-track stereo master. Master mixes must be consistent with the tastes of artists and producers while at the same time recognizing each recording's intended audience. A final mix may be saved on digital audiotape (DAT), compact disk (CD), or reel-to-reel tape (still preferred by a few engineers for its analog warmth). From these two-track stereo masters, multiple copies can be duplicated for distribution. In some instances, a stereo mix recorded as an AIFF or WAV audio file may be uploaded to a server via the Internet where it can be held for subsequent broadcast or duplication.

Engineers tend to develop their own personal approaches to mixing stereo masters. Many find that a good way to start is to establish a mix for rhythm tracks first. This will often be the bass guitar, percussion, and rhythm guitar or perhaps keyboard. These instruments really serve as the foundation of the mix, and getting their mix in proper shape often requires the most time and effort. Once these tracks are mixed, properly panned, and equalized, then the lead tracks can be layered on top. It can be useful to conceptualize a mix as made of layers with background elements at the bottom and foreground tracks at the top. Layering can also be envisioned in frequencies, some of which occupy the lowest bass ranges, some at the low mid-ranges, and others at the high mid-ranges and upper treble ranges.

Mixing is basically the construction of a sonic image by adjusting each track's level, location in the stereo field, and equalization. Certain tracks, such as lead vocals or lead instruments, should be placed in the mix at a higher level because of their importance to the overall sound. Normally, producers prefer to place these tracks at or near the center of the stereo field. To "open up" a mix, and to allow individual instruments to be heard distinctly, other tracks may be panned to the left and right channels in varying degrees. Some might be panned **hard left**—meaning to only the left channel—some slightly left of center, others right of center, and still others **hard right**. This technique spreads the instruments across the stereo field, making the sound more spacious and giving it realism.

If there are more than a few tracks, there is a danger of creating a cluttered final mix. Attention to the frequency content of each track can help avoid this problem. In contemporary popular music, many of the recorded tracks are likely to contain similar frequency ranges. Mid-range frequencies tend to become overpowering and blur together indistinguishably. These frequencies are supplied by keyboards, guitars and vocals, and, to some extent, drums. To remedy this, each track can be given a slightly different EQ emphasis while maintaining separate pan positions.

Mixing conventions have arisen that producers and engineers tend to follow, a few of which have evolved from practices developed during the era of vinyl recordings. Musical artists still frequently refer to their projects as "records," even though finished products are rarely available on vinyl recordings anymore. As noted previously, listeners generally expect lead vocals to be placed in the center

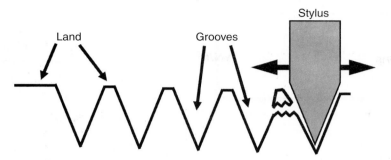

FIGURE 15.6   Overdriving the recording stylus can break the land between grooves on a disk recording and cause skipping to the adjacent groove on playback.

of the mix (panned equally in the left and right channels) and out front, or louder than the remainder of instruments and voices.

As noted, bass guitar and kick drum are also usually centered in the stereo field. If a bass is centered the signal will be present in phase in both channels making it sound louder and fuller. This practice is partly connected with habits that have survived from vinyl recording days. In cutting a disk on a cutting lathe, a heated stylus plows a groove into a smooth plastic platter. Lateral movements of the stylus record the signal as the disk rotates beneath it. Low frequencies cause wide excursions of the stylus. If the bass is panned hard left or right, exaggerated excursions of the stylus can break through the groove wall and into an adjacent groove. This is illustrated in Figure 15.6. If this happens, record **skip** is sure to occur when playing the record.

USING EFFECTS

It is during a mix-down that equalization enhancing certain frequencies or de-emphasizing others is finally applied. An engineer's ability to make such adjustments is contingent on the presence of those frequencies on tracks to begin with. This can be only achieved by careful engineering during tracking sessions. Effects may also be added to alter the recording's sonic impression. Because of the close miking technique normally employed in tracking sessions, reverberation is frequently added to selected tracks. This fosters an aural illusion of the greater spaciousness of a more natural environment.

Most studios provide a selection of outboard effects such as digital reverbs, delays, compressors, and multi-effects processors. To add an outboard effect to a mix, the console's effects sends or auxiliary sends are used to route a portion of the signal to the processor. Each channel incorporates sends or buses for this purpose. But to add the processed signal to the mix, the producer must also **return** the affected signal to the console, as illustrated in Figure 15.7. This is accomplished

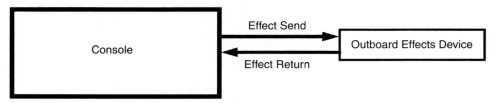

FIGURE 15.7 An outboard effects loop sends a dry signal from the mixing console to the processing device, and the signal with effects is then returned to the console to be mixed with other signals.

using the effects return bus. The returns allow the **wet**, or modified signals, to be mixed in proper proportion with the dry, or unaffected signals, and sent on to the two-track master.

Novice engineers sometimes find this process confusing. On increasing an effects send level, they may expect to hear the resultant effect on the track, not realizing that a return must be set up as well. This way of managing effects processing by means of an **effects loop** has a long history. In former times, reverberation processing employed an **acoustic reverb chamber**, as illustrated in Figure 15.8. The method was simple: (1) send a dry signal to a speaker in a reverberant room, (2) pick up that signal and all its reflections with a microphone placed in that room, and (3) return that wet signal to the console where it was mixed with the dry signal.

A good rule of thumb for the use of effects should be "less is more." The single error most frequently made in the control room by apprentice engineers is the heavy-handed use of effects. When the effect is sufficiently pronounced in the mix to call attention to itself, then it almost certainly is too much. The aim in adding delay processing is to create a sense of the space in which the performance took place, never to have the listener say "nice reverb!"

Most studios include **patch bays** that offer a simple means to break a signal path and to redirect it to an alternative destination. Patch bays significantly expand options for routing signals to and from outboard effects devices. In studios with dozens of outboard processors but employing a console having just a few effects or auxiliary buses, the patch bay affords easy access to all processors' inputs. Most patch bays are **normalized**, meaning that the signal is routed to a predefined destination unless a patch cable is inserted into the signal path at the patch bay. The use of a cable at the patch bay allows the signal to be detoured to processing units. The top illustration in Figure 15.9 presents a normalled signal flowing to an effect device labeled 1. The bottom illustration presents a patch cable inserted into the signal path, diverting the signal to effect device 2. In Figure 15.10 the photograph shows a patch bay using bantam or TT (tiny telephone) connectors at the top; the bottom patch bay uses quarter-inch connectors.

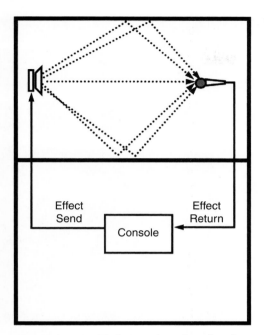

FIGURE 15.8   Early reverberation chambers were simply reverberant rooms outfitted with loudspeakers and microphones. The dry signal was sent to the loudspeaker and the microphone returned the processed signal.

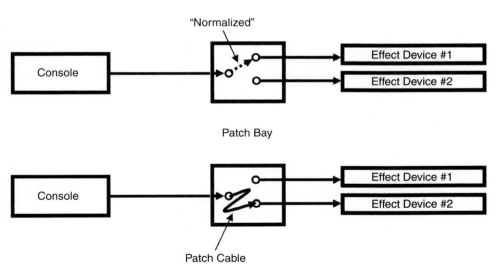

FIGURE 15.9   When no patch cable is inserted into the normalized patch bay, as in the drawing at the top, signals flow to device number one. When the patch cable is inserted, as at the bottom, the signal path is interrupted so that signals are diverted to device number two.

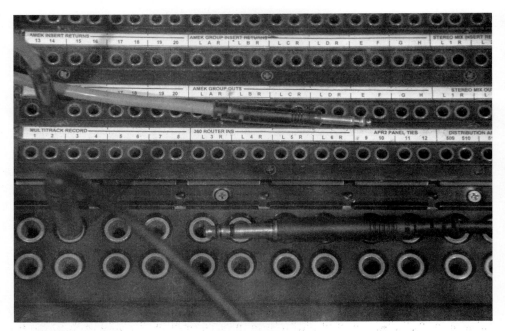

FIGURE 15.10   Patch bay and patch cables: Bantam or TT at top, quarter-inch at bottom.

So much processing capability is available in modern studios that a stereo master might hardly resemble the original performance. In such cases, one might say that the processing gear becomes, in effect, part of the musical ensemble. In fact, audio facilities afford performers so much creative latitude that some recorded studio performances can be difficult for artists to reproduce in a live performance.

RECORD KEEPING

Multiple takes, files, reels or disks, and their track assignments are a continual source of confusion for master mix engineers. To keep things organized, multi-track engineers make use of a **track sheet** to simplify record keeping and ensure efficient use of time in the mix-down facility. Track sheets record important information for each session: date, engineer, artist, the instrument(s) and effect(s) recorded on each track, and, if an analog master, tape stock. Mixing information may be added, including levels and panning assignments for each track. Consoles with automation and software-based digital audio workstations eliminate the need for much of this information because these details are retained by the system electronically. Labeling master recordings is equally important. Whether stored on CD, DVD, flash storage, or tape, all masters should be tagged with the same information—date, engineer, artist, number of cuts or takes, and, possibly, microphone placement information.

CHAPTER SIXTEEN

SOUND FOR PICTURES

The discussion in the following section presents procedures that are specific to audio production for video and film. Sound for pictures generally falls into three distinct categories: dialogue, sound effects, and music. In each of these categories, special practices have emerged for capturing, editing, mixing, and distributing products.

DIALOGUE FOR FILM AND VIDEO

Catching dialogue for television broadcasts can be as simple as using a familiar but conspicuous handheld microphone. This technique, common in interviewing and newsgathering, customarily employs a dynamic microphone with an omnidirectional polar pattern. The omnidirectional pattern is forgiving of errors in the mic's sound source coverage, as interviewers tend to concentrate on what is being said, not on the mic's aim. This advantage comes with a serious drawback—the omnidirectional mic is subject to high ambient noise pickup.

Broadcast audio technicians frequently make use of the **lavaliere microphone** described in Chapter 4, Microphones. The "lav" is available in either wired or wireless configurations, dynamic or condenser types, and with varying frequency response characteristics. Wireless lavalieres can be used to capture dialogue in motion pictures, particularly in scenes that place the camera at a substantial distance from talent. An example might be a shot of talent walking along a crowded sidewalk made with a camera at a fixed location, or a scene in which a performer enters an automobile and drives away while speaking to another person.

A more common tool for capturing film dialogue, especially in closeups and medium shots, is the shotgun microphone. The shotgun, or hypercardioid microphone, exhibits an extremely narrow polar pattern. The use of a **fish pole**—a long, lightweight fiberglass shaft—to mount a shotgun mic permits sound technicians to remain off-camera even while holding their microphones close to performers. Shotguns are attached to fish poles using shock mounts to cut handling noises, and they need to be outfitted with a windscreen to reduce rumble and other noises from air movements. This technique is described in detail in Chapter 14, Field Recording. The shotgun mic may be held above or below the performer, just beyond camera range. The exact positioning of pole-mounted mics should be chosen to minimize ambient noise pickup.

The boundary microphone is another option for dialogue use. This microphone is known for its ability to minimize off-mike sounds. Its hemispheric pickup pattern has been employed to successfully record dialogue in automobile interiors and other enclosed spaces that do not lend themselves to the use of boom microphones.

Single-camera productions are commonly taped or filmed out of sequence, and this can produce uneven audio quality among shots. Irregularities arise from changes in background noise and from timbre changes caused by shot-to-shot variations in camera and microphone setups. Location shoots often encounter unwelcome ambient noise that cannot be managed by microphone selection or placement. Motion-picture location-sound recordists commonly attack this problem by covering the noise with environmental sound. It is possible to do this by recording pure "room tone," the naturally occurring ambient sound present within a location environment. This is then mixed into the sound track at a low level to provide acoustic consistency in the final edit.

Sound captured during filming may alternatively be **wild sound**, *production sound,* or *direct sound.* Although direct sound may be employed in documentary productions, it is rarely retained in the final release prints of dramatic or narrative films. This is true of both location and studio footage. Film sets can be as noisy as location environments. Camera motors, arc lights, crew movements, motorized equipment, and the like generate dialogue tracks that contain excessive background noises. To overcome this problem, the practice of filming **MOS**, or "mit out sound" became the norm. (Allegedly, early German-born film directors pronounced "without sound" that way.) In this method, dialogue is rerecorded in a studio setting where audio quality is more readily maintained and the sound is added in postproduction.

To replace dialogue a process known as **looping** may be employed, the term acquired from the way voice replacement is done. Originally, a piece of film containing a single shot of action was made into a loop and projected on the screen in a sound studio. The same shot would be played over and over again as the film looped through the projector. The talent could repeatedly voice their lines to match exactly the film image. The best match of sound with the moving image would be selected for the film's final release print.

Even though technology changes have produced quieter motion picture sets, much of film dialogue is still rerecorded. Looping, however, has given way to **automated dialogue replacement**, or ADR. ADR allows performers to hear the prerecorded sound dialogue recordings as well as see the film clips. This allows better matching of sound and screen images. ADR may even be used to replace individual words or lines for reasons such as script changes, dubbing for international distribution, or reediting for broadcast and airline release.

SOUND EFFECTS FOR THE SCREEN

Like dialogue, **sound effects** are seldom captured live during filming or videotaping; they are usually created and added in postproduction. Sound effects can be either **synchronous**—that is, synchronized to on-screen action—or asynchronous, not matched to any on-screen event. If production budgets are small, sound effects from commercially available compact disks can be employed. But in major motion pictures and network television shows, sound effects are individually created to better match on-screen action. This process is known as Foleying, a term applied in recognition of pioneering film sound recordist Jack Foley, who specialized in the technique. **Foley** sound effects specialists have a distinctly different skill set than other audio production workers, and their artistry can have a significant impact on the realism of finished works. Major film and television productions include subtle sounds, which audiences will not be conscious of, but which add naturalness to the viewing experience.

Foley sounds tend to fall into three categories: **moves**, **footsteps**, and **specific sound effects**; so, in films, three separate Foley tracks are created for each one of these. Moves effects are the sounds of rustling clothing, hand movements like scratching or rubbing, and other sorts of environmental sounds that viewers may not notice unless they are absent. Foley artists create moves sound effects by manipulating fabrics or by patting or rubbing hands in synchrony with action on the screen.

It has been said that a good Foley studio looks like a junk shop. A wide assortment of items must be close at hand for simulation of sounds ranging from a gun (created with a staple gun) to birds flying (made by moving a pair of gloves)—all sounds of the sort that comprise what are called "specifics" effects. Specifics include the sounds of crashes, crunches, punches, squeaks, bangs, and all the other sounds associated with unambiguous on-screen actions.

A conspicuous feature of Foley studios is the **pit**. Pits are insulated enclosures, typically about three feet by four feet, that contain a variety of surfaces on which Foley artists walk or step to simulate on-screen characters' footsteps. Each character in every scene will be assigned a separate footstep track. This may seem to give unnecessary attention to detail, but sounds of characters' walking, running, or scuffling and shuffling have a powerful influence on a scene's believability. Foley walkers match their footsteps to on-screen action, but they do it while walking in place because pits are only a few feet wide, and the microphone is fixed on a boom not far above the pit's surface. Among the surfaces usually available in studio pits are hardwood, stone, carpet, grass, gravel, mud, sand, and collection of other materials chosen to approximate surface textures of different movie locations. To ensure exact sound matches with footfalls, Foley studios may also keep large collections of different shoe types.

Sound designers and Foley engineers take pride in their ability to catch subtleties in sound produced in natural environments and to duplicate these effects artificially. By listening critically to everyday sounds such as handling, rattling, shaking, rubbing, and striking objects, Foley technicians are able to catalogue sounds for future association with film action.

Some film sound effects are pure invention. Obvious examples can be found in genres such as science fiction, horror, and fantasy films. The task of creating sounds of intergalactic spacecraft, futuristic weapons, and alien creatures is not one of duplication but wholly of imagination. Effects like these can be created either by synthetic generation or by combining and manipulating real sounds. For example, in creating a sound palette for *Star Wars,* Ben Burtt famously made use of tricks such as pitch changing, speed changing, and reversing naturally occurring sounds. Burtt captured scores of natural sounds, modifying and mixing them to make the extraterrestrial effects needed. To create the sound of laser weapons, Burtt started with a recording of a hammer blow to a broadcast antenna's high-tension steel guy wire. The reverberation of the sharp impact traveling along the cable was heavily processed and combined with other noises to create the movie's signature sound.

Apart from these unusual sonic fabrications, techniques used by Foley engineers are not so different from ones used in other productions. One exception is that, more than in music recording, Foley productions are likely to use large amounts of equalization to exaggerate the sonic qualities of effects. Otherwise, Foley sessions start with a good-quality microphone, usually condensers, placed within three feet of sound sources. Once good-quality audio is captured, normal practices of creating clean, aesthetically tasteful master recordings are followed.

MUSIC FOR FILM AND VIDEO

Arguably the most important element of a film or video soundtrack is its music. Film producers use a musical **score** to establish a movie's mood and to suggest the time and place of a story's setting. Because the music plays a supporting role it is also known as **incidental music**. In the days of silent films, long before movies had dialogue or sound effects, they had music. These films were accompanied mainly by piano or organ performances, but occasionally small orchestras were used to provide musical backing. Music quickly became an essential part of the presentation of movies, and producers experimented with other methods of accompaniment. One approach to film sound made use of phonograph disks that were loosely synchronized with films. Both the Photophone by RCA and the Vitaphone by Western Electric used methods like this. When magnetic and **optical soundtracks** became available, for the first time films could achieve genuine synchronization. This finally made it possible for films to talk.

The means and efforts invested in a musical accompaniment can vary greatly, ranging from utilization of prerecorded commercial music libraries at one extreme to the composition and recording of an original score performed by a large studio orchestra at the other. A recent trend favors incorporation of hit popular music as part of the soundtrack.

How films integrate music differs too. In some instances, music may be written and recorded beforehand and the footage cut to match the music, but in most major motion pictures music is recorded after the film or video is edited into near final form. By building the audio and visual elements in this order, musicians can view scenes while performing so that music cues precisely match on-screen action.

Music for film may be conceptualized in one of two ways: scores or songs. Although it is true that the art of traditional orchestral film scoring has not yet been lost, it is costly and beyond the budgets of many independent filmmakers. In the classic big Hollywood movie production model, composers scored films through the preproduction and production stages of filming. During postproduction, musical scenes were performed and recorded on a **scoring stage**.

A scoring stage is an expansive facility having ample space for a symphony orchestra and a projection screen for the conductor, composers, and musicians to view footage to which they will match their performance. As in Foley looping, scenes are projected while the orchestra performs the score. It is the responsibility of the conductor to keep the musical performance in step with on-screen action.

Contemporary film and television scoring has adapted to changing musical styles and technologies. The rising popularity of affordable digital technology has made the art of film scoring accessible to more filmmakers. Whereas traditional screen composers such as Elmer Bernstein and John Williams wrote lush arrangements for orchestras, other composers, including Mike Post, Danny Elfman, and Don Davis, now create soundtracks using synthesizers, sequencers, and **samplers**. Today, independent and student filmmakers, and even local television producers, can afford the expense of original musical scores.

The incorporation of hit recordings and other popular music in films has been practiced since at least the 1950s. In the beginning, this was quite a departure from the usual method of custom-made music. In 1955, *Blackboard Jungle* introduced the tune "Rock Around the Clock," performed by Bill Haley and the Comets. The song went to the top of the pop music charts, perhaps boosted by its inclusion in the film. The following year, Hollywood's first full-length rock-and-roll film, *Rock Around the Clock,* also featuring Bill Haley and the Comets, was released and included several of the group's previously recorded songs. In this movie, the Comets lip-synched musical numbers, a technique that became a standard practice over the next half century. Movies exploiting prerecorded songs have continued to gain popularity, and their presence has boosted sales of soundtrack recordings. By the year 2000, record producers such as T-Bone Burnett were given key roles on film production teams because of their knowledge of recorded music and their skill in matching tunes with movie narratives.

The technology for precision synchronization of sound and image has been greatly simplified and made more affordable. Early methods of film scoring were complicated, requiring specialized equipment and production facilities. Synchronization made use of a **magnetic soundtrack**, often on magnetic full-coat film, 35-millimeter motion picture film coated with the same **oxide** used on magnetic recording tape. Complete sound tracks—including dialogue, sound effects, and music—would only be mixed as they were transferred to magnetic full coat. Throughout the editing of movies, sound and pictures were cut separately because they existed on two independent reels of film. Synchronization of sound and image was possible by physically aligning **sprocket holes** on the film and the full coat. Editing was accomplished on vertical or horizontal flatbed editors of the sort shown in Figure 16.1. Film laboratories matched sound and film footage, converted magnetic to optical sound, and printed both onto the film's release prints.

FIGURE 16.1   A Moviola film editor allows motion picture film to be viewed in synchronization with sound played back from magnetic full coat film. Film editors use this system to edit picture and sound elements into final form.

SYNCHRONIZING BY TIME CODE

The ease and precision of film and sound synchronization improved markedly in the 1960s with the introduction of **time code. SMPTE** (Society of Motion Picture and Television Engineers) time code is a stable and precise locator code that can be magnetically or optically recorded on both the picture and sound tracks. Every frame of film or video is identified by a unique code, hence SMPTE time code is said to be **frame accurate**, or accurate to within one frame. This amounts to 1/30 second for video or 1/24 second in film. Once coded this way, sound and images can be separately processed and then rejoined at a later time by simply matching their codes. SMPTE time code information can be visually displayed in a window on a video screen in the format of "hour:minute:second:frame." If a time code window displays 01:23:15:06, it means that the frame currently displayed is located one hour, twenty-three minutes, fifteen seconds, and six frames into footage from the start of coding (Figure 16.2).

SMPTE time code generators and readers may be used to synchronize all sorts of film, video, and audio equipment. A generator creates a stream of time

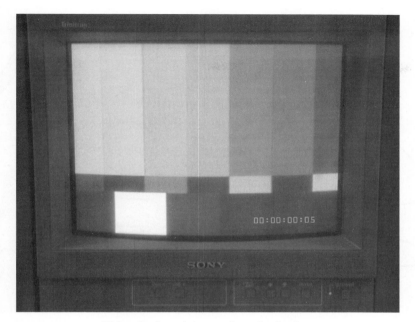

FIGURE 16.2   Video display of time code and color bars.

codes, supplying it to a recorder for storage alongside matching frames. When provided with codes on playback, a reader converts them to time and frame information that can be used to display or to lock pieces of equipment into synchronization. Once locked together by SMPTE time code, the stream of codes will regulate the speed of playback devices.

This capability opens a number of possibilities. Sound tracks can be recorded and edited separately, and as long as the time code is carefully preserved, the audio can be later synched up with the images, accurate to one frame. This also allows replacement of soundtracks that, by means of time code, can be exactly matched with original footage. This ability has profoundly affected the production of music videos, as performers can lip-synch to their master audio recordings while cameras film their performance. As cameras record new visuals and time code is transferred from the original musical performance, artists act out their performance for the cameras. Using time codes, even multiple takes of visuals can be edited together while maintaining exact synchronization. Figure 16.3 illustrates a typical layout for time code linkage between a film camera and an audio recorder. This setup permits prerecorded time coded audio to play back through a speaker, allowing talent to lip-synch while a camera captures film images synchronized by time code. In addition, in real-time shoots, a system like this can be used to record time code simultaneously onto film and audiotape.

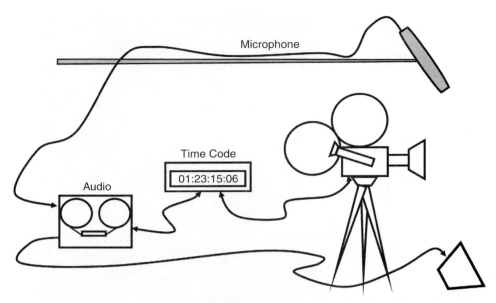

FIGURE 16.3 Time code permits video equipment to sychronize film, video, and audio recordings.

DIGITAL SYSTEMS FOR THE SCREEN

Computer-based recording and nonlinear editing systems have revolutionized the way moving images are fashioned into movies and television programs. They have also dramatically advanced the state of the art in sound for pictures. Avid Media Composer was one of those tools that ushered in the new digital era. In digital nonlinear editing, film is digitized for computer storage, after which all editing takes place in specialized workstations. This has become the industry's standard method for major motion picture production. The availability of digital, high-definition, twenty-four-frame progressive-scan video cameras capable of a wide-screen-aspect ratio (16:9), makes it possible for directors to abandon expensive, fragile, and cumbersome 35 mm film stock in favor of digital video throughout the entire production process. For example, George Lucas shot *Star Wars Episode II— Attack of the Clones* entirely on video.

Most professional nonlinear editing systems incorporate multitrack audio capabilities. This feature allows multiple audio tracks to be recorded, edited, and mixed simultaneously with viewing and editing of images. Similar to processes used in digital audio-only workstations, audio tracks can be moved along a time-line so that audio events are perfectly matched with visual cues. Figure 16.4 shows the on-screen display from a simple video editing application. Three stereo audio tracks are provided and designated as voice, music, and sound effects. More so-

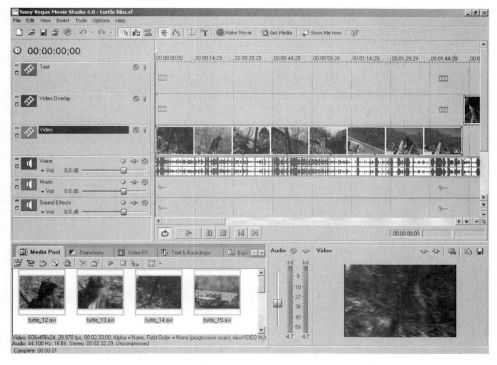

FIGURE 16.4   The screen display of a nonlinear video editor, showing multiple tracks of video and audio.

phisticated editors provide dozens of audio tracks onto which multiple layers of audio material may be recorded.

SPECIALIZED FILM AUDIO SYSTEMS

Another noteworthy development in sound for film and video has been **surround sound**. Although movies as early as Disney's *Fantasia* incorporated multichannel sound, it was introduced to film on a major scale in the 1970s with stereo sound. Tape noise-reduction pioneer Ray Dolby and his company, Dolby Laboratories, played a leading role in developing present multichannel film sound technology. Multichannel movies grew from stereo to include rear speakers that are standard in surround systems. In the early 1980s Dolby Laboratories introduced Dolby Stereo, utilizing optical sound printed onto the film. Interestingly it actually offered not two but four channels: left front, right front, center, and surround. The Dolby surround sound system employed a scheme that placed four channels of audio on the two optical soundtracks available on 35 mm film. The surround channel was merely meant to be a rear speaker, but it was put to creative use by filmmakers who could actively pan sounds to the surround channel, thus creating an illusion of movement.

Dolby Laboratories seized on this idea to further develop audio surround technology. The company offered **Dolby 5.1**, increasing surround channels to two with the intention that they would be placed on the left and right sides of audiences instead of the rear. Thus, the five channels are: left front, right front, center, left surround, and right surround. The ".1" refers to a subwoofer channel. **Dolby 6.1** further expands the surround channels by adding a rear center channel: left front, right front, center, left surround, rear surround, rear center, and, of course, the .1 subwoofer. Figure 16.5 illustrates typical speaker placement for both Dolby 5.1 theaters, with the additional channel for Dolby 6.1 indicated by the box having dashed lines.

A curious aspect of surround sound is in the way the system delivers sound to theater audiences in the Dolby DTS (Digital Theater System). Optically encoding multiple digital-sound audio tracks on film presented many problems so, harkening back to the early Photophone RCA phonograph sound system, Dolby Laboratories adopted compact disks synchronized to film projectors for audio playback. Naturally, modern technologies have made synchronization much more accurate and reliable than when projectionists had to approximate a match between phonograph records and pictures.

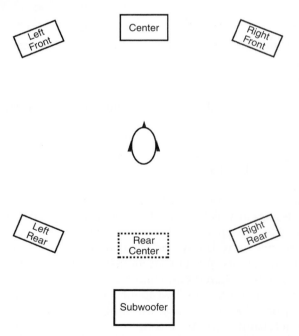

FIGURE 16.5   Typical surround sound loudspeaker placement for 5.1 or 6.1 surround sound.

THX is frequently thought of as another film sound system, but THX is much different than Dolby. THX was developed by George Lucas' Lucasfilm Ltd. to ensure a consistent high-quality audio for the film-viewing experience. THX design parameters extend to the establishment of guidelines for ambient noise levels in the theater, reverberation, distortion, and equalization. The THX system also specifies ideal sight lines, viewing angles, and image brightness. For a venue to be THX certified, the theater must adhere to the lengthy and detailed set of standards laid out by the Lucasfilm THX Division. Central to that certification is the leasing of proprietary THX crossover units combined with other items of equipment such as amplifiers, speakers, and other equipment identified as THX approved. When all these conditions are met, theaters may display the THX logo to indicate that viewers will hear sounds as close to those intended by the producer as can be achieved. Several manufacturers offer THX equipment and systems for professional and consumer markets.

The THX system has enjoyed popularity among sound for film engineers because it presents them with predictable sound playback environments. If an engineer mixes a soundtrack on a THX certified system in a THX certified control room, the variables that usually give rise to undesirable sound aberrations along the audio presentation chain will be tightly controlled.

VIDEO GAME AUDIO

Video games have emerged as an important entertainment medium within recent decades, and games are heavily dependent on sound effects and music. The earliest commercially distributed video games were "Pong" type games, first marketed in 1972. Sounds in these games were just crude electronic beeps. By the 1980s, adventure games began to feature computer-generated music. Synthesizer chips in game consoles of that era relied on programmed musical sequences contained in game cartridges to generate sounds. Limitations in the console and chip capabilities restricted music to short sequences only. As games grew in sophistication, audio accompaniments for them have had to keep pace.

Contemporary CD-ROM-based game consoles allow game designers to complement on-screen action with recorded music, sound effects, and realistic dialogue. These sounds function in much the same way that film sound supports and matches up with on-screen images. Digital compression of audio recordings is used to conserve storage requirements so that a single game disk may easily contain over sixty minutes of audio sequences. Games may still use synthesizer-based music and background sounds, but it is possible to incorporate fully orchestrated scores using recording methods familiar in the film industry. Everything in between these extremes is possible too—chamber music, choral pieces, blues, rock, and jazz. Composers of game music usually write their music in a way that lends itself to looping and to easy transition from piece to piece. Music in games tends to shift in tempo and mood as action moves through different levels or phases. As in films, game music may be conceived as either scores or songs. And just as in films, game designers may license an already popular hit song for inclusion in their products. The gaming industry has even reached a sufficient stage of development that some game designers are able contract with popular music artists to collaborate in the production of original musical pieces for new game projects.

■ ■ ■ ■ ■

CHAPTER SEVENTEEN

EDITING, SWEETENING, AND POSTPRODUCTION

Audio and video production projects are usually conceived as consisting of three distinct phases: **preproduction** (planning), **production**, and **postproduction**. In this section, postproduction for audio is described. For multitrack recording projects, postproduction processing is most often called **mastering**, while for film and video projects, the audio postproduction is more often referred to as **sweetening**. The terms are actually somewhat interchangeable, and broadly refer to the process of making final adjustments in gain and equalization, and correcting phase inconsistencies. Postproduction may also include editing as well as related tasks needed to ready a project for distribution.

MASTERING TECHNIQUES

To some, mastering is a mysterious and incompletely understood process, and audio engineers who work in other phases of production often admit they do not really know all that is involved in mastering. Even a cursory survey of mastering studios and mastering houses quickly reveals that their engineers tend to have individual approaches to the art of mastering. In fact, not all engineers use the term *master* consistently. Some refer to the final two-track mix as the master. In the following pages, the term *master* is used to describe the final edited, equalized, and processed product that will be sent to a pressing plant or disk-stamping house. In the definition used in this text, a *master* is the recording from which the consumer product is duplicated. Masters may be created from the two-track mix, but a two-track mix is not considered the master until it has undergone the final stages of processing.

In multitrack music recording, mastering procedures have changed over the years, evolving with the emergence of digital technologies and with the introduction of new means of distribution. At an earlier time, mastering meant the process of preparing a master recording for submission to a record-pressing plant or tape-duplication house, working from a final two-track analog tape mix. To do this, the master tapes were optimized for signal levels. This was accomplished by adjusting the signal sent to the mastering recorder so that the highest levels would just begin to saturate the tape and be on the verge of producing distortion of the waveform peaks. The master would then be recorded with levels reduced only enough to ensure no distortion. This same result could be achieved by using a limiter, and some engineers used them to ensure accuracy in levels. However, many purists sought to avoid the addition of further processing in the signal chain, preferring instead to manage levels manually.

In traditional mastering methods, individual audiotape cuts were edited for length and spooled together with each piece separated by leader tape on the master reel. The length of the splicing tape between the individual tape segments dictated the period of silence between cuts on the finished album. Editing for length might also entail fading a song at its end, or physically cutting passages from the tape with a razor blade as described in Chapter 8, Audio Editors and Editing. Mastering engineers would record a series of tones at the head of a master reel of tape. An oscillator generating a sine wave at 100 Hz, then 1,000 Hz, and finally 10,000 Hz would be recorded at 0 VU on the mastering recorder. When these tones were played back at the record-pressing plant, it allowed the engineer there to adjust head alignment and reproduction equalization on the playback machine to match it with the machine used to record the master. The reel of tape would be packaged "tails out," that is, with the end of the tape on the outside of the reel. This was done to reduce print-through from the outside layer of tape to inner layers, which would produce faint echoes. After all of this, the master tape would be ready to cut a master disk as the first step in the mass production of vinyl disks. The master tape would then be played back and the output sent to the input of the cutting lathe to record the master acetate disk.

Nowadays, mastering generally necessitates the preparation of a two-track mix that has been recorded in any one of a number of different analog or digital formats. Ideally, the mix should be recorded at the highest sampling rate and bit depth available. That is to say, the recording should make use of an 88.2 kHz rate and 24-bit samples if possible. Use of these parameters ensures the greatest possible signal-to-noise ratio for the recording. Mastering still includes editing cuts for length, arranging the sequence of cuts on the master recording, equalization, and signal processing. Deciding on the sequencing of material is the producer's responsibility; the mastering engineer merely follows instructions provided on delivery of the two-track recordings.

The output of a multitrack recording project is typically a two-track mix, although projects intended for distribution on DVD are frequently mixed in 5.1 surround sound. Regardless of whether a final mix is recorded to a two-track analog reel-to-reel tape, a DAT, a compact disk, or other storage format (engineers and producers will have their own favorite recording medium), the master recording will likely need additional treatment before being mass-produced on CD, vinyl, or tape for distribution to consumers.

Different engineers tend to approach the mastering process in distinctly different ways. Indeed, specialists in this field can be somewhat secretive about their personal techniques for successful mastering. Even so, most include the same steps at some point in their production sequence. Mastering today is, of course, carried out on digital workstations, and this allows mastering engineers to visually edit the waveforms if needed. Each cut will be trimmed to eliminate any small amount of "dead" material that might be present before and after the recorded piece. As always, the cuts in the project will be placed in the sequence in which they will appear on the final product. Then the amount of time between each cut can be fine-tuned to fix the length of the pause between pieces. Mastering software allows the songs to cross fade from one to the next, pause for a few seconds, or go instantly from cut to cut as desired.

Adjusting levels not only involves ensuring that levels are consistent from one cut to the next, but also that the overall levels on the disk are maximized. As anyone who listens to compact disks should have noticed, not all disks are created with equal levels. Some seem "hotter," exhibiting higher average sound levels. Although still a matter of personal preference and production style, most artists and producers prefer that their disks sound "hot," "punchy," and generally louder than other recordings. This is especially the case for recordings aimed at radio play, where artists seek to have their work stand out from other recordings in the steady stream of music on air.

Most workstations used in mastering provide processing plug-ins for normalizing. Normalizing is a little like companding (compressing and expanding), where a maximum level is set by the user, and the entire digital wave is either raised or lowered so that peaks rise to, but do not exceed, that preset level. In normalizing, *relative* levels within the waveform are maintained. That is, low-level passages are not raised to peak levels but will continue to be held at the same level as compared to the loud passages. Normalizing can produce a more consistent and natural sounding result than limiting or compressing, and is an easier way to ensure maximum signal level without danger of overdriving digital circuits.

Exceeding maximum recording levels in a digital tape results in extremely unpleasant digital distortion. Indeed, if digital signals on a compact disk are pushed beyond their maximum limits, players may simply reject the disk and respond with an error message after spin-up, or there may be severe audio dropouts. In the analog era, boosting levels during mastering meant pushing levels to the tape's **saturation** point. Exceeding that saturation point might introduce distortion, but analog tape was more forgiving than digital media and the resulting sound was not so disagreeable.

Mastering engineers frequently contend that producing a disk with a "hot" sound is contingent on achieving phase consistency on the tracks. Phasing errors can be introduced at several different stages in the recording process. Phase problems can easily come from wiring irregularities. If hot (+) and cold (−) wires are reversed on one end of a cable, the phase of signals they carry will be inverted by 180 degrees. More commonly, phase errors occur when sound arrives at two microphones at slightly different points in time, as detailed in Chapter 4, Microphones.

In addition, a recording's phase integrity can be defeated by the performance of musicians. This phenomenon can be seen in the speaker cone movement of amplified instruments. Imagine that a bass note moves the cone of the woofer forward. As the phase passes its peak and begins to decay, the cone will begin its backward movement toward its neutral point. If the attack of another low frequency note begins during this backward travel, the rising phase will push the cone to move forward again in opposition to the cone's backward travel. Engineers, producers and musicians use expressions such as "playing in the groove," to describe performers playing in exact unison so that slight phase errors do not creep in. Although recording software does provide tools to adjust notes along the timeline to achieve perfect synchronization, such corrections need to be accomplished at the time of mix-down to the two-track master.

In addition to altering sound characteristics, phase differentials affect levels too. Two identical signals that are exactly 180 degrees out of phase will cancel each other. Other smaller phase differentials will reduce the level by lesser amounts and distort their combined frequency content. Occasionally phase problems of various sorts can be corrected in mastering, but more often they are impossible to repair. The best solution is to strive to deliver a clean product to the mastering house, rather than adopting a "fix it later" mentality.

EQUALIZATION IN MASTERING

Equalization has always been a major part of the mastering process. In the vinyl-disk era, recorded signals required *preemphasis* when transferred to a phonograph record. To improve system signal-to-noise ratios, preemphasis would be applied to boost high frequencies and to cut low frequencies at the time of disk mastering. On playback, deemphasis would reverse the recorded equalization, with highs being cut and lows boosted. The end result was a flat frequency response, with the audio raised above the noise floor. Before 1954, each record manufacturer used their own preemphasis scheme, but after that date the Recording Industry Association of America (RIAA) published a set of recommended equalization specifications, the **RIAA Curve** and this became a worldwide standard. Consumer stereo equipment manufactured after 1960 employs built-in RIAA deemphasis circuitry. A set of inputs labeled **phono** normally provides proper equalization for phonograph playback.

Equalization problems can be introduced during mastering or in the mix-down process if attention is not paid to the principle of equal loudness. The principle of equal loudness states that the human ear is most sensitive to mid-range frequencies, but that as sound levels increase the perception of high- and low-frequency sounds improves. The louder the signal gets, the better hearing becomes at high and low frequencies. Following the implications of this relationship, it can be seen that if monitoring occurs at too high a level during mixing, there will be a tendency for the engineer to cut equalization at the ends of the audio spectrum. When that mix is heard at a normal level, it will have a prominent mid-range and will appear to lack a full-frequency range. A parallel problem results when monitoring levels are excessively low during mix-down, as the tendency will be to boost highs and lows too much. This results in a mix that lacks mid-range.

To guard against equalization mistakes during mastering, engineers are advised to reference several types of monitors. Most mastering engineers tend to employ a favorite monitor with which they have a high degree of familiarity, and in which they have confidence in accuracy and frequency response. In spite of this, hearing a final master played through speakers of several sizes and designs helps assure engineers that the final product will present good definition and a balance across the audio spectrum. Doing this also helps ensure the acceptability of the final sound in a variety of listening conditions.

PROCESSING IN MASTERING

Some mastering engineers place great importance on specialized signal processing such as compressing, **de-essing**, or more esoteric proprietary effects like the **stereo enhancer**, the **Aural Exciter**, or the **Sonic Maximizer**. De-essers are available as software plug-ins for software-based editors, and as freestanding boxes that can be placed in effects racks and added to the signal chain. It should be kept in mind that not all engineers agree on the utility of specialized processors of this sort. Proprietary processors, including enhancers and exciters, often combine multiple effects and corrections in a single box or plug-in.

De-essers, such as Drawmer's MX-50, are essentially "smart" equalizer/compressors. When excessively high frequencies are detected in the range where human voice **sibilance** occurs (between 800 Hz to 8 kHz), the de-esser will lower (compress) those levels by a user-defined amount. Frequencies above 8 kHz can be left unaffected, in order to preserve the high-frequency content of other sources. While the de-esser can be used during tracking, it is touted as a repair device that fixes oversibilance in the mix. It is also a tool commonly used in the mastering phase of production.

According to BBE Sound, the Sonic Maximizer accomplishes two processes: correcting phase inconsistencies in higher frequencies and boosting the high and low frequencies. The design is based on a logic that as frequency increases, so does impedance. (Note that this is only true in circuits that have an inductive characteristic.) According to this theory, higher frequencies suffer from a phase shift, so that higher frequencies are said to arrive at listeners' ears later than lower frequencies. The Sonic Maximizer is offered as a corrective for this. The bass/treble boost component of the processor is derived from the previously discussed principle of equal loudness.

Aphex Systems manufactures the Aural Exciter, which is marketed as an "enhancer." Product literature for the Aural Exciter explains that the device improves detail, clarity, and imaging. The Aural Exciter extends frequency response by restoring harmonics and overtones lost in the recording process, and may even compensate for limited frequency response from a poor microphone. Aphex's intelligent design claims only to affect portions of the signal that require enhancement, by extending the frequency content of tracks with complex harmonic content, but not simple signals (e.g., sine waves) without harmonics. Basically, the Aural Exciter is said to create harmonics in user-defined frequency bands, adding them to the signal.

As suggested near the beginning of this chapter, *sweetening* is a term used in film and video productions to describe the general use of effects and equalization, the balancing of signal levels, and the reduction of noise. As the term implies, sweetening is the use of processing to make recorded sound more pleasing. Sweetening may involve adding a bit of reverberation to an otherwise dry sound narration track. It may involve balancing levels and equalization among separate shots from location shooting, so that the background sound is consistent throughout a visual sequence. Sweetening may even require replacing dialogue, by using ADR techniques described in Chapter 16, Sound for Pictures.

MAKING COPIES FROM MASTERS

As in other branches of the audio field, analog disk-recording technology has evolved radically over the years. Whereas Edison's original cylinders were made of foil, and then were later replaced by wax, ceramic, and other brittle materials, and early disks were made of equally fragile shellac, contemporary disks are made of vinyl polymer **polyvinyl chloride**, a plastic that is a highly flexible yet resilient material. And whereas the earliest recordings were original recordings cut directly onto a cylinder or disk, and then sold directly to consumers, disks purchased by today's listeners are mass-produced through a stamping process.

Contemporary recordings begin with a lacquer-covered aluminum disk onto which grooves are cut by the heated stylus of a record-cutting lathe like the one shown in Figure 17.1. The lacquer-covered disk (sometimes termed an acetate be-

FIGURE 17.1   A disk lathe is equipped with a heated stylus to cut grooves on lacquer-coated aluminum platters.

cause the lacquer is based on an acetate plastic compound) is then electroplated with nickel. The nickel master is formed from the lacquer disk. The master can be thought of as a kind of inverse or negative copy of the acetate, with ridges instead of grooves. From this master a metal **mother disk** is produced that is a positive once more, with grooves instead of ridges. From the mother, **stampers** are cast. Stampers are metal negatives (ridges) that are fitted on the record press, one for each side of the record. Figure 17.2 shows a press, ready for insertion of stampers.

Heated balls of vinyl called **pucks**, as shown in Figure 17.3, are placed into a press, along with record labels properly located in the stampers. The record press closes and squeezes the hot, partially melted vinyl into the flat grooved record that, when placed into a sleeve and cover, can be sold to consumers. Interestingly, although vinyl used in record production can be any color, or even transparent, only a few records are manufactured using anything but black vinyl materials. Raw vinyl pellets ready for conversion into pucks are shown in Figure 17.4. Figure 17.5 shows an extruder that heats and forms the vinyl into a puck.

The manufacturing of compact disks somewhat resembles production processes employed for vinyl records. CD manufacturing begins with the creation of highly polished glass master disks that are coated with a photoresistive chemical layer.

FIGURE 17.2   A press is used for production of vinyl records by pressing softened vinyl between metal stampers.

FIGURE 17.3 A vinyl puck before insertion into a disk press.

FIGURE 17.4 Raw vinyl pellets prior to heating.

FIGURE 17.5   An extruder converts pellets into vinyl pucks by melting the plastic and forming it into viscous balls.

Audio material is recorded on the disk by a laser that burns pits into the photo resist. This glass master disk is then plated with a layer of metal. A second layer of metal plating is applied to the glass master and then peeled off to create three to six "fathers." The fathers are negatives, with bumps rather than pits. From the fathers the same type of electroplating process is used to create three to six positives called "metal mothers." The metal mothers are then used to create stampers, which are used in an injection molding process to manufacture CDs for the marketplace.

Unlike vinyl records that record from the outside to the center of the disk, compact disks are read from the center outward, and CDs spin in a direction opposite to vinyl disks (counterclockwise). Also, compact disks do not spin at a constant rate, as records do. Proper data conversion from digital to analog requires that the number of bits that are being processed by the CD player remains constant and that the speed of conversion or **transfer rate** remains steady. Because the circumference of the disk is much greater on the outside of the disk than at the center, one rotation at the disk's edge yields much more data. As a result, the rotation of the CD must slow as it is read from inside to outside, because each rotation contains increasingly more data bits.

Mastering for CD and DVD release requires a different approach to control the dynamic range than does mastering of either vinyl or audiocassette recordings. Analog recording media, particularly vinyl, have a much more restricted signal-to-noise ratio, no more than 60 dB at best. In contrast, digital recordings at 16-bit 44.1 kHz can easily achieve better than a 90 dB signal-to-noise ratio. This means that analog releases may need substantial compression to shrink the audio's dynamics into a suitable range. For some kinds of recordings, including most popular music, the dynamic is not likely to be great anyhow, so the treatment at mastering time may not need to be very aggressive. In any case, digital recordings provide enough dynamic range to necessitate little compression of material under most circumstances.

CONCLUDING THOUGHTS

Mastering engineers face greater challenges today than they did before the 1990s because increasingly, work delivered to them is produced under less-than-ideal conditions. Because of the growing popularity of low-cost professional quality equipment, many project or home studios have been built in facilities that lack effective acoustic treatment. Productions from these studios frequently need corrective processing to repair or hide defects. Moreover, musicians and producers who lack experience and expertise in quality recording work often operate these studios, and their productions frequently need mastering processes that correct uneven levels and clean up the finished product.

In order to get the most from the mastering process, a producer must deliver clean, quality recordings, properly mixed, and ready to be given final touches for the production plant. The mastering engineer should not be counted on to fix problems introduced during the tracking or mix-down phases. Engineers may be able to do it, but this will only slow down the process and add to costs. At the very least, conventional methods of mastering and sweetening give producers an additional set of trained ears to hear to their projects before release. As an independent mastering engineer prepares the final product, each piece will be given an unbiased and careful examination, with special attention to levels and equalization. Because the tracking engineer, mixing engineer, and artist are party to the production from its earliest stages, they may develop preconceptions that affect their ability to evaluate critically finished recordings. The mastering engineer is subject to none of these influences and can bring a fresh perspective to the final production phase.

POSTSCRIPT

BE KIND TO YOUR EARS

It should be obvious that human hearing is the most valuable thing an audio professional possesses; a producer's ears are essential and irreplaceable instruments that should be treated with great care. And yet, hearing loss and damage across the world is at an all-time high. Many audiologists attribute the epidemic of hearing problems to high volume levels delivered through "ear buds" that many listeners use with personal MP3 players. The high sound pressure levels produced within the ear canal can be much greater than that created by loudspeakers or headphones. These high sound levels, coupled with the large amount of time that many users spend listening to personal players, are raising concerns among those involved with hearing health. According to the BBC, an Australian study found that about a quarter of persons who use personal music systems such as iPods were listening at levels that risked permanent damage. Some MP3 players can provide 105 dB of output, but authorities recognize 80 dB as sufficient to cause some degree of short-term loss, and prolonged exposure can produce permanent loss. One sign of excessive sound levels is a ringing or buzzing in the ears after listening, a clue that some injury has been caused to hearing.

THE NEW TECHNOLOGY LANDSCAPE

The proliferation of digital transmission and storage devices has brought costs of sophisticated systems down to levels that make them accessible to the semiprofessional and serious amateur segments of the sound recording and music production fields. In practical terms, this has meant that home recordists and amateur musicians now have access to systems that can capture, edit, store, and reproduce audio signals exceeding the sonic clarity of even commercial recording studios in the late twentieth century.

Of course, just as in the analog tape era, the quality of the signal is only as good as the weakest link in the recording chain. But with a moderately well-equipped desktop computer, a good sound card, and some moderately priced microphones, home engineers can create genuinely professional-quality recordings. It is literally true that the specifications and track capabilities of many amateur set-ups far exceed those of the world-class studios of the 1970s and 1980s.

Obviously this ability to create and reproduce sonically flawless recordings and copies of recordings has also spawned a conflict between the recording industry and a certain group of consumers. One can easily make copies of recorded music that are of the same high quality as the originals. Presently, as everyone knows, these files can be easily shared with other music fans via the Internet. So in this closing section, the authors lay out the important legal and ethical issues that producers face throughout their careers.

INTELLECTUAL PROPERTY RIGHTS

Some may feel that commercially recorded music is unreasonably expensive, and they may use this as a justification for sharing recordings. People like this almost certainly would have a different opinion if they were relying on revenues from recording sales for their livelihood. People who are lucky enough to earn a living as audio producers are likely to become passionate about discouraging music pirating. Aspiring audio professionals who read this text are moving to the other side of the producer–consumer equation. When the answer to the question "Whom does it hurt?" is you, you are less likely to burn that copy for a friend.

Consider the intricacies of current laws on **intellectual property rights**. Our ideas about intellectual property rights predate the founding of the United States of America, and can be traced back to eighteenth-century British law. The notion that an individual could own intellectual creations, or ideas, is grounded in the U.S. Constitution. Depending on the nature of a creation, a person can claim ownership through either patents or trademarks.

In a nutshell, the copyright law presently in force states that you can own, and profit from, your ideas as long as you live. After your death, your heirs may continue to profit from your creations for an additional seventy years, through renewals. After a copyright has expired, a creative work falls into the public domain, and may be used freely by anyone. Because copyright laws have been revised periodically throughout U.S. history, about all one can say for certain is that works created before December 31, 1922, are in the public domain.

For most eligible creations, January 1, 1978, is a critical date; anything produced after that date will receive protection for lifetime plus a fixed term of years, plus a bonus of additional years as set by Congress. Determining the lifespan of creations produced before that date is too complex to detail here. Assuming that one wishes to register one's own newly created works for copyright, the following provisions of the law apply to those works. The interpretation provided by the U.S. Copyright Office is that a work "is automatically protected from the moment of its creation and is ordinarily given a term enduring for the author's life plus an additional 70 years after the author's death. In the case of 'a joint work prepared by two or more authors who did not work for hire,' the term lasts for 70 years after the last surviving author's death. For works made for hire, and for anonymous and pseudonymous works (unless the author's identity is revealed in Copyright Office records), the duration of copyright will be 95 years from publication or 120 years from creation, whichever is shorter." (Accessed on the World Wide Web 20 March 2007 at www.copyright.gov/circs/circ1.html.)

Many recent legal cases have involved instances in which an artist or composer has been accused of appropriating a melody line from an existing song. In defending these lawsuits, some composers claim that any influence was subconscious, and therefore not malicious.

Things are further complicated in some types of contemporary music by **sampling** and looping techniques. Sampling involves digitally capturing a segment of a recorded performance. Looping takes that sample and then repeats it indefinitely. In an infamous case involving sampling, rap artist Vanilla Ice was found guilty of copyright infringement for sampling Queen's "Under Pressure," then using the basic bass/drum riff as the foundation for his recording "Ice Ice Baby." Even though the segment of the Queen song that was sampled was only about one measure, the courts found that it was so fundamental to the work that it acted as the song's "hook" and could therefore be protected by copyright law.

In today's litigious society, it is prudent to obtain permission from the composer before sampling some element of an existing work—even just a "beat"—to incorporate into a new recording as a loop. Avoiding the subconscious influence of preexisting works in composing is not so easy. As some composers would argue, there are a finite number of notes and chords available, and so some duplication is practically unavoidable. It is easy to notice similarities in chord progressions and melodies in portions of many popular songs.

THE STATE OF THE FIELD

Another result of the proliferation of affordable recording equipment, and the boom in "made-at-home" recordings, is a fundamental change in the structure of the recording industry. This has created an enormous challenge to major studios and major labels. World-class recording studios, with their high overheads, are forced to charge hundreds of dollars per hour in order to generate sufficient revenues to service their debt. This has encouraged more artists to build their own home studios as a money-saving tactic, and to gain more control over their creative environments and products. Some industry prognosticators foresee a decline in the number of large full-service recording studios like those that dominated the recording industry during the second half of the twentieth century.

The concentration of ownership that generally pervades the media industries is fully present in the audio recording field. A handful of multinational corporations are responsible for the bulk of the recordings sold in the United States. One result of this dominance has been a backlash among some consumers and artists. These consumers tend to regard the concentration of ownership as a root cause of skyrocketing CD prices, and artists blame the big companies for declines in their personal earnings. These attitudes have undoubtedly helped fuel battles over Internet file sharing, the decline in CD sales, and the impressive growth of small independent and artist-owned labels. With such dynamics at work, changes in the recording industry's landscape are sure to continue in the years ahead.

It might be argued that another outcome of the explosion in availability of inexpensive high-quality recording technology has been a concomitant decline in quality of the recorded products available to consumers. Even though their sound fidelity may be good, the production, engineering, and musicianship may not be. Indeed, the market has become flooded with poorly engineered live recordings, flat-sounding studio recordings, and good recordings of bad material. This is very much like the trend afflicting the Internet, which is flooded with poorly researched, poorly written, and poorly designed "junk" material. The recorded music industry's filters have become weakened as it transitions into a cottage industry. Powerful but discriminating **A&R** officials at the major studios can be circumvented if artists release their vanity recordings independently. So while some might herald the democratization of the recorded-music field as a great step forward in the free expression of ideas, some others might decry it as the end of quality controls.

You, the audio professional, will face decisions that pose both ethical and legal dilemmas. The authors suggest that you consider those questions thoughtfully, with the guiding principles being is it fair, is it legal, is it the best I can do with the tools that I have? Will you contribute to the advancement of the recording arts and sciences?

GLOSSARY

A&R Artists and repertoire. In the record industry, A&R staff members have the job of finding and developing recording talent.

absorbers Absorbers, also sometimes called diffusers, are designed to capture and attenuate sound. In the studio, absorbers are often used to provide acoustic isolation between instruments and/or vocalists.

AC Alternating current. An electrical current that reverses its polarity periodically. One example is household electrical current that has a frequency of 60 hertz (Hz) in North and South America or 50 Hz in most other countries.

acoustic energy Energy conveyed by sound waves.

acoustic foam A spongelike porous material, in which the open cells "trap" air molecule movement, and thus inhibit reflections and subsequent reverberation. Acoustic foams, like all such porous surface treatments having small openings, generally offer absorption limited to high frequencies. So while acoustic foam can attenuate high frequencies that contribute to a "bright" reverberant sound, it does little to eliminate booming at lower frequencies.

acoustic isolation Techniques used to prevent the transfer of acoustic energy from one physical space to another.

acoustic reverb chamber An analog system for producing reverberation by routing audio through a room designed to produce abundant sound wave reflections.

acoustic suspension speaker A loudspeaker design in which the speaker driver is housed in an airtight enclosure. The trapped air in the box reduces the degree of forward and backward movement of the loudspeaker through damping. A type of "infinite baffle" enclosure.

acoustic treatment Materials attached to the walls or other surfaces within a studio to reduce problematic room resonances or to make the studio space less reflective or "live."

acoustics The name given to the scientific study of sound and sound energy.

active loudspeaker systems A powered loudspeaker system that contains its own amplifier.

ADAT Alesis digital audio tape. A system developed by Alesis Corporation for recording digital audio on Super VHS videocassettes.

ADC Analog-to-digital converter.

AGP Accelerated graphic port. This bus system is used on computer motherboards to connect with high-speed graphics cards.

AIFF One of several algorithms for recording audio digitally. Developed for use on Apple Computer platforms, it uses the file extension .aiff.

air check A recording made from an on-air monitor, or receiver tuned to a broadcast station. Used for archiving, demonstration purposes, or quality control of programming.

air studio   The studio used for direct, on-air broadcasts, in contrast to a studio used for recording and editing material for later broadcast. In radio, the primary air studio and the control room are often one and the same. Audio signals from the air studio/control room travel directly to a station's transmitter.

algorithm   A mathematical formula or a step-by-step solution for a problem.

alternating current   See *AC.*

ambience   A quality arising from the presence of background sounds that give a natural impression of the sound environment.

amplifier   A device that increases the power or voltage level of an electronic signal.

amplitude   A measure of the power or level of variables such as an electrical signal or a sound wave.

amplitude modulation   A radio and television transmission technique in which the amplitude or power of the carrier wave changes in direct proportion to the level of the modulating signal. Employed for radio broadcasting in the medium wave or standard broadcast band. In the NTSC television system, the video carrier is amplitude modulated with the audio subcarrier being frequency modulated. Abbreviated as AM.

analog   A continuous electronic signal that varies in both frequency and amplitude according to sound waves that it represents. The resulting electrical signal is analogous to the original sound wave. Compare with *digital.*

anti-aliasing filter   A filter that removes frequencies above the Nyquist frequency. See *Nyquist frequency.*

ASCAP   American Society of Composers, Authors, and Publishers, a major performance rights organization. Like BMI, ASCAP licenses musical compositions on behalf of its member songwriters and publishers for broadcast and public performance. ASCAP then distributes money it collects to its member songwriters and publishers according to a predetermined formula.

assigned   In audio, selected by means of a switch, usually refers to routing of channel outputs to subgroup, group, or stereo inputs.

asynchronous   In sound for video, a description of sounds not synchronized or matched with any on-screen event. See also *synchronous.*

ATRAC   Adaptive transform acoustic coding. An audio compression codec used in Sony's MiniDisc (MD) systems.

attack time   A measure of how quickly a compressor or limiter responds to changes in input audio levels.

attenuation   A reduction in the amplitude, or level, of a variable such as an electrical signal or sound wave.

attenuator   A device that produces attenuation. See *attenuation.*

Audio Engineering Society   A professional organization devoted to advancement of audio and audio standards. Abbreviated AES.

audio frequencies   The range of sound frequencies, from low bass to high treble, which correspond to optimal human hearing—nominally, 20 Hz to 20,000 Hz. Few people can hear all frequencies in this range because of limitations brought on by age, sustained exposure to loud sounds, and other physical factors.

Aural Exciter A proprietary audio processing system marketed by Aphex Corporation.

automated dialogue replacement (ADR) A system used to record film dialogue by allowing performers to hear prerecorded sound.

automation Devices that permit equipment to operate unattended and without repeated intervention by humans.

auxiliary A term used for a general-purpose send.

auxiliary send See *auxiliary buses.*

auxiliary buses In audio mixers, auxiliary buses are output channels provided in addition to the primary output channels to afford greater flexibility in routing signals.

averaging meter A meter that provides an approximate average indication of peak level. An averaging meter does not provide instantaneous readings for peak voltages of a waveform.

baffle The surface to which a loudspeaker driver is attached. Its function is to prevent acoustic energy produced at the rear of a loudspeaker cone from combining with acoustic energy emitted at the front of the loudspeaker cone. A baffle is thus a barrier to sound waves.

balanced Audio connections in which negative and positive inputs are kept separate from ground circuits.

banana plug A heavy single-conductor connector. Used in situations where electrical currents may be large.

band pass A filter that has defined low-frequency and high-frequency cutoff points, thereby allowing frequencies between those two points to pass through without attenuation. Frequencies above and below the cutoff points of a band pass filter will be reduced in level.

bandwidth The range of frequencies contained between defined upper and lower limits.

basic tracking In multitrack recording, the session in which initial tracks are recorded. These serve as a foundation and guide for tracks added later.

basket In a dynamic loudspeaker, the metal frame on which all other components are mounted.

bass Low frequencies, commonly associated with musical instruments such as bass guitars, tympani, and tubas. Also, the notes below middle C on a piano.

bass reflex speaker A design in which the loudspeaker enclosure has a hole at its front that permits acoustic energy to escape from the box. Properly constructed, the phase of the acoustic energy released through the hole, or "port," matches that of the primary acoustic energy coming from the front of the loudspeaker. The result is that the two sound waves reinforce each other at bass frequencies.

Bel The basic unit for expressing the ratio of two power levels. The decibel, one-tenth of a Bel, is more often used.

bi-amplification An arrangement in which the tweeter and the woofer of a speaker system have their own separate amplifiers.

bias A high-frequency audio signal generated by an oscillator within an analog magnetic recorder. It is combined with audio prior to recording. The level of the bias signal should be properly adjusted for the tape formulation being used. AC bias moves the audio signals away from the zero point to linear portions of the magnetic transfer curve. The bias signal's frequency is ordinarily at least five times higher than the highest audio frequency to be recorded.

bidirectional   Having two lobes of sensitivity 180 degrees apart, as in a bidirectional pattern.

bidirectional pattern   A figure-eight microphone response pattern. The microphone picks up sounds best from the two directions represented by the lobes of the figure-eight at the front and back.

binary   In mathematics, a numbering system having two possible values, conventionally 0 and 1. See *bit*.

binaural dummy head   See *binaural microphones*.

binaural microphones   Stereo systems employing two microphones placed on a dummy head at positions corresponding to human ears. Binaural microphones can achieve a dramatic audio realism.

binder   The adhesive or gluelike substance that attaches the oxide coating of magnetic recording tape to its backing.

binding posts   Connectors used to attach bare wires to a piece of equipment using finger-tightened screw mechanisms. Also called five-way binding posts.

bioacoustics   The study of sound communication and sound location processes.

bit   A binary digit. Eight bits commonly constitute one byte. Hence, a byte would be a binary number made up of eight zeros and ones. Computer memory and storage capabilities are commonly measured in kilo-, mega-, and gigabytes.

Blumlein technique   A stereo setup in which the diaphragms of two bidirectional microphones are placed very close to each other but at a relative angle of 90 degrees.

BMI   Broadcast Music Incorporated. One of the two major performance rights organizations representing music composers and publishers in the United States. The other major performance rights organization is ASCAP. A third, smaller organization is SESAC.

BNC   A latching twist-on connector used mainly in RF equipment.

boom   A long arm, usually made of metal and used for suspending a microphone near sound sources.

bouncing   Mixing two or more audio tracks to an available empty track in order to free up the original tracks for further multitrack recording.

boundary-effect microphone   A microphone having a hemispherical pattern that is designed for placement on a tabletop, wall, or other flat surface. Minimizes off-mike effects.

broadcasting   Radio or television transmissions intended for reception by the general public.

buffer   A digital storage system used to ensure continuous regular flow of data. A buffer is placed between an input and an output where it functions as a reservoir, managing the outward flow of data and helping to reduce interruptions in the data stream.

bus   A pathway for the flow of digital data or electronic signals.

bus assign See *assigned.*

byte A set of digital data typically consisting of eight bits.

calibration Adjusting a meter so that its readout accurately indicates values. Also, adjusting recording equipment so that the VU meter readings on each recorder are precisely matched. For example, with a steady 1,000 Hz tone set at 0 VU on a mixer, the input VU meters of recorders connected to the output of the mixer should also be adjusted to read 0 VU.

capacitance The ability of an electronic device to store an electrostatic charge.

capstan The metal shaft connected to the transport motor of a tape deck. The spinning capstan provides the primary force pulling tape past the deck's erase, record, and playback heads. To move the tape, a pinch roller presses the tape against the capstan.

carbon microphone A microphone type that uses a pickup element made of a cup containing tightly packed carbon granules pressed against a thin metal diaphragm. As the compression of carbon varies by action of acoustic waves, the mic produces an electrical signal analogous to incoming sounds. Carbon microphones have high noise levels and poor frequency response.

cardioid See *cardioid pattern.*

cardioid pattern A directional microphone pickup pattern so named because it resembles a heart shape.

carrier The radio-frequency signal that serves to carry signals modulated on it from transmitter to receiver.

cartridge An audio tape having its ends spliced together to form a continuous loop contained in a plastic shell or cartridge. Used in radio prior to introduction of computer-based digital cart systems.

cassette Usually a plastic boxlike housing containing audio or video tape attached to feed and take-up reels and associated hardware. Cassettes provide protection for delicate recording tape.

CD Compact disk. An optical audio disk containing digital data representing voice, music, or other material recorded and reproduced by means of optical technology.

CD quality The quality of sound equivalent to digital audio recordings made with a sampling rate of 44.1 kHz and a bit depth of 16 bits.

CD-ROM Compact disk read-only memory. A compact disk containing digital data such as computer data and software.

channel A common pathway within a mixer carrying a single signal. Depending on the type of console, it may be an input channel intended for collecting an input signal or it might be a bus carrying a signal to other destinations within the mixer.

channel strips Modules that each contain circuitry for a mixer channel.

chorus effect A special effect producing a sound somewhat like a chorus of voices or instruments. It is commonly provided in keyboard synthesizers and both hardware and software effects processors.

choruser Device used to produce chorus effects.

circuit A closed electrical loop, from a Latin word that means, roughly, a "circle" or to "go around." When a circuit is opened, electrical current cannot flow.

click track A tape track, or the DAW equivalent, on which are recorded audible clicks that provide a tempo reference for the musicians, somewhat like a metronome.

clipping Truncating the peaks of an audio signal. The result is a distorted waveform.

close miking Placement of a microphone close to a sound source. Close miking allows reduction of sounds from unwanted sources nearby and decreases the pickup of room reverberations.

coaxial Sharing a common axis.

coaxial cable An electrical cable containing a center conductor surrounded by an insulator, which is itself surrounded by a wire mesh or foil shield. The entire cable has an outer layer of plastic or rubber insulation. The electrical elements share a common axis, and so they are coaxial.

coaxial loudspeaker A speaker assembly made of a tweeter placed in front of—and on axis with—a larger, woofer driver.

codec Compress/decompress or code/decode. A computer algorithm for recording and playback of audio or video data.

coil In electronics, an inductor made of a conductor arranged in a coiled shape so that the magnetic field of each turn adds to the device's combined magnetic field.

comb filter A filter characterized by a response curve that resembles the teeth of a comb. Frequencies passed by the comb filter alternate with frequencies that are sharply attenuated, giving the filter its special graphical representation.

companding The application of both compression and expansion within a single system.

comping Short for *compiling* or *composite*. A technique in which the best sections of several takes of a recorded performance are combined into a single composite track.

compression A reduction in dynamic range, as in a compressor. Also, a reduction of the size of a digital data file, either by use of a lossy or a lossless codec.

compressor Hardware or software that shrinks, or "compresses," the dynamic range of audio material.

condenser microphone A microphone design that uses a capacitor, or "condenser," as its pickup element. Condenser microphones typically have the most faithful reproduction, though they tend to be highly sensitive and, thus, prone to overloading.

conductors Materials through which an electrical current may flow. Metals such as copper, silver, aluminum, and gold are good conductors.

cone A common term for a dynamic loudspeaker's paper or plastic sound-producing diaphragm.

console One of several names commonly applied to the control desk or mixer.

control room The on-air control area of a radio station.

control room monitor A loudspeaker system located in the control room that allows audio engineers to monitor ongoing production work.

control surface Devices used in digital workstations to provide tactile control over operating parameters.

crossover network A filter network used in loudspeaker systems to route the appropriate frequencies to their respective drivers. In a loudspeaker system employing two drivers, the high frequencies would be routed to the tweeter and the low frequencies to the woofer.

crossover points The frequencies at which crossover networks divide signals for loudspeaker drivers.

crosstalk Interaction between electronic circuits in which a signal from one is induced in another.

cue To prepare audio for playback or editing. On broadcast consoles, a cue channel allows the operator to listen to audio sources off-air and prepare them for broadcast. Also, in some mixers, a bus used for auxiliary purposes.

cutting lathe The machine used to carve grooves containing audio into blank disks.

cycle In an AC signal, a complete interval beginning at zero and rising to a positive peak, through zero again and through a negative peak, and finally back to zero.

DAC Digital-to-analog converter.

damping Reduction of the amplitude of a wave or signal.

DARS Digital audio radio service. A satellite technology to deliver multiple channels of radio programming to home users.

DASH Digital audio stationary head.

DAT Digital audio tape.

data compression The process of shrinking the size of audio or video data files or data streams.

DBS Direct broadcast satellite.

dbx noise reduction One of several companding systems employed to improve signal-to-noise ratios in analog recordings.

DC Direct current. The flow of electrical current in one direction only, e.g., as in a battery-operated device.

DCC Digital compact cassette. A largely outmoded digital recording technology developed by Philips in the late 1980s. Philips Corporation designed DCC players to be compatible with analogue audiocassettes.

dead In audio, either the absence of sound or the absence of reflected sound, as in a studio environment.

Decca Tree A variation on the spaced pair stereo miking technique in which three omnidirectional microphones are placed just behind and above the orchestra conductor's position, about six-and-one-half feet above the floor.

decibel (dB) One-tenth of a Bel, commonly abbreviated and referred to as dB. The decibel is defined as the smallest change in loudness that the trained ear can detect. In fact, the average listener can usually detect changes in loudness no smaller that 3 decibels. The decibel is a relative measure, as it actually describes differences in loudness. Named in honor of telephone inventor Alexander Graham Bell.

decouple To interrupt the physical link between two systems.

deemphasis See *preemphasis*.

de-essing Technique for reducing sibilant sounds in voices.

delay A short interval echo.

desk See *console*.

detent A click stop.

diaphragm A thin element in a microphone used to capture air molecule movements caused by sound waves. This motion is then transferred to pickup devices.

dielectric A material that does not normally pass electrical current—an insulator—that is placed between the plates of a capacitor.

diffusers A type of surface treatment used to inhibit reflections, causing waves to scatter in different directions.

digital Technology based on measurements of a waveform made at precise intervals. These measurements are recorded as binary numbers. Compare with analog.

digital audio stationary head See *DASH*.

digital audio tape See *DAT*.

digital audio workstation (DAW) A workstation based upon computer technology that is used for recording, editing, and processing audio.

digital cartridge systems Computer-based systems in which digital audio files are recorded and played back somewhat in the manner of traditional analog cartridges.

digital compact cassette See *DCC*.

digital console A type of audio mixer that performs all processing in the digital domain.

digital signal processing (DSP) A means of modifying characteristics of audio signals through the application of algorithms that alter the values of digital samples. By signal processing, it is possible to create effects such as reverberation, compression, limiting, and equalization.

digital snake A type of snake intended to carry digital signals. See *snake*.

DIN Deutsche Institut für Normung (German Institute for Standards). In audio, a type of connector used widely in MIDI interfaces.

direct current See *DC*.

direct to disk As the name implies, direct to disk is a recording made directly to a disk without use of any intermediate recording.

distortion The undesirable alteration of an audio waveform. Generally, distortion is undesirable because it makes the resulting sound less faithful to the original sound, but distortion can be used creatively. Rock guitarists often use distortion processes to change the characteristic sounds of their instruments.

distributed sound system A technique for delivering sound to a large number of loudspeakers spread across a large venue. A transformer steps up the voltage at the power amplifier output, and at each speaker a transformer reduces voltage to an appropriate level again. This approach cuts power losses across the system.

Dolby 5.1; Dolby 6.1 Surround sound systems.

Dolby noise reduction system Noise reduction systems developed by Ray Dolby and his company. These schemes all employ companding techniques that boost high frequencies on recording and cut them on playback.

dominant frequency The fundamental frequency within a sound wave.

doubling Recording the same sound twice then combining the two by mixing.

download To retrieve a computer file from a remote server.

dry Without processing.

DTRS Digital tape-recording system. A system developed by Tascam Corporation for recording digital audio on Hi-8 video tape.

duty cycle The length of time that a piece of electronic equipment may be used continuously without the risk of overheating or other types of damage.

DVD Digital video disk. Also known as digital versatile disk, to reflect uses other than video.

dynamic loudspeakers Loudspeakers that rely on the production of sound by a current flowing through a voice coil attached to a diaphragm or cone. Movements of the cone are caused by the interaction of the magnetic fields of the coil and a permanent magnet.

dynamic microphone A type of microphone that uses a moving voice coil within a magnetic field to generate an electrical analog of sound waves.

dynamic range The difference between the softest and the loudest levels in a performance or a recording. In a piece of equipment, dynamic range is defined by its signal-to-noise ratio.

ear buds Earphones that fit into the ear canals.

early reflections The first reflections of sound within an enclosed space.

earth See *ground*.

editing The selection and organization of recorded sound elements. Editing may involve removal of undesired portions of recorded audio.

editor A person who edits audio. Also, the software and/or hardware used for editing.

EEPROM Electrically erasable programmable read-only memory. An electronic component that contains permanent software (commonly called firmware) that can be rewritten electronically.

effects The alteration of any characteristic of sound such as dynamic range, equalization, or delay.

effects loop A complete processing scheme for routing that includes effects send and effects return.

eight-to-fourteen modulation A means of digital optical recording that cuts effects of data loss on playback as a result of dust and scratches.

eight-track cartridge A consumer music format popular in the 1970s based on a long loop of tape contained within a plastic cassette. Recordings on the tape could be played continuously.

eighth-inch mini plug A single-pin connector, one-eighth inch in diameter.

electret condenser microphone A condenser microphone that uses an electret to charge its diaphragm. Electrets are devices that maintain a permanent electrical charge. Even though the electret charges the microphone's diaphragm, it still requires power for its internal preamplifier, usually supplied by penlight cells inside the microphone case.

electromagnetic Possessing both electrical and magnetic properties. A flow of electrons in a conductor induces a magnetic field surrounding it. Also, moving an electrical conductor through a magnetic field generates a voltage in the conductor.

electromotive force (EMF) Force that moves electrons in a circuit. Also known as difference of potential, electron potential, or colloquially, voltage.

electron A negatively charged subatomic particle. Electrons are visualized as locked in orbit around an atom's nucleus.

electrostatic loudspeaker A loudspeaker design that incorporates a large, thin conductive diaphragm panel sandwiched between two electrically charged grids. Conceptually, electrostatic loudspeakers are similar to condenser microphones.

energy The capacity to perform work.

EOM End of Message. A tone or digital cue that indicates the ending of a commercial, jingle, song, or other recorded audio material. Used in automated radio systems to start play of the next item in a program list.

equalization (EQ) The process of increasing or decreasing the amplitude of selected bands of audio frequencies.

equalizer A device that produces equalization.

erase head A magnetic tape recorder head that scrambles polarities of magnetic particles on magnetic tape, effectively erasing any previous recording.

fader A variable resistor, or potentiometer, used to adjust the gain of an audio source.

far field Monitors placed sufficiently far from the listener to allow reverberation to become a significant part of the sound heard.

FCC Federal Communications Commission. Regulatory agency for broadcasting in the United States.

feed reel A plastic or metal reel from which tape feeds past a tape deck's heads to a take-up reel.

feedback In an amplified circuit, the return of a portion of the output to the input. Feedback in an audio amplifier can lead to self-oscillation, producing loud, ear-piercing squeals or growls.

feedback controller A tool that combines the functions of a real-time analyzer, graphic equalizer, and compressor/limiter to reduce gain in frequency bands in which self-oscillation caused by feedback is detected.

Fidelipac A leading manufacturer of broadcast audio cartridges.

field recording A recording not made in a recording studio.

file server A central networked computer on which files are stored so that they can be made available for use or downloading by others.

filter An electronic component that allows selected bands of frequencies to pass while attenuating other frequencies.

FireWire Apple Computer's term for high-speed digital data transfer between a computer and a peripheral using the protocols of IEEE 1394. Sony Corporation uses the name "iLink" for connections that use the same specification.

fish pole A long pole used to mount a microphone so that it can be brought close to sound sources. Commonly used when shooting video.

flanger A device that produces flanging.

flanging effect A comb filter effect that sweeps across a range of audio frequencies.

flat In audio, having a response that is equal across the audio spectrum.

Fletcher-Munson curves The Fletcher-Munson curves describe the human ear's frequency response, as standardized through a series of widely administered listening tests. Human hearing is most sensitive to frequencies between 1 kHz and 6 kHz, and the ability to hear frequencies below 500 Hz and above 7 kHz progressively declines. See principle of equal loudness.

flutter In analog tape recording, small but rapid variations in tape speed that produce slight repetitive pitch changes on playback. The more smoothly that tape moves across a tape deck's heads, the less flutter and wow will result. See *wow*.

flutter echo Rapid recurring echoes in an acoustically live sound environment constructed of parallel walls or other surfaces.

flying fader Motorized faders that move according to programming provided by automated consoles.

FOH Front-of-house. Refers to space in the live performance setting where the audience is located.

foil reverb A mechanical reverberation device that employs a large sheet of foil held under tension. Transducers are attached at varied locations on the foil to generate and pickup audio signals.

foldback bus The bus used to mix audio sources to be sent to performers so that each one of them can hear the whole group as a properly balanced sound.

foldback monitors Onstage monitor speakers that face toward performers, allowing them to hear a mix of their performance. Also, in radio, an audio intercom system linking control room console operators with performers broadcasting from a remote site.

Foley A motion picture technique for adding postproduction sound effects matched to screen action. Named for its originator, Jack Foley.

footprint In satellites, the area of the earth covered by a communication satellite.

footsteps In Foley audio, the creation of footfall sounds matched to on-screen performance.

frame accurate Video and film systems having timing accuracy of one frame, $\frac{1}{30}$ second in video or $\frac{1}{24}$ second in film.

free electrons Electrons that may be easily displaced from their atomic orbits.

frequency In audio, the number of cycles per second of a repeated electronic or acoustic wave. Measured in hertz (Hz). The fewer cycles that repeat per second, the lower will be the sound's perceived pitch.

frequency masking See *masking*.

frequency modulation In frequency modulation, the transmitter carrier's frequency varies according to the amplitude of an audio signal. The amplitude of the radio wave remains essentially constant. FM was developed by radio pioneer Edwin Armstrong. Abbreviated as FM.

frequency response The frequency spectrum that a device is capable of reproducing or processing with specified signal levels.

front-of-house See *FOH*.

FTP File transfer protocol. An set of codes for uploading and downloading computer data files via a network or Internet connection.

full-track A single magnetic audio track that occupies nearly the entire width of a quarter-inch-wide audio tape. See *track*.

fundamental The principal or predominant frequency of a signal.

gain A measure of amplification, commonly expressed in decibels.

gain stage A circuit that produces gain. A gain stage has an output level that is greater than the input level.

gain trimmer A potentiometer used to adjust preamplifier gain.

generation loss Losses in audio quality, especially signal-to-noise ratio that occurs when making analog copies of a recording.

geostationary See *geosynchronous*.

geosynchronous A satellite that orbits the earth at the same rate of speed as the earth's rotation, making the satellite appear stationary to observers on earth.

giraffe boom A long mic boom mounted on a wheeled tripod.

graphic equalizer An equalizer that incorporates a series of bandpass filters each associated with a fader assigned to a specific band of frequencies, typically one-third, one-half or a full octave wide. The positions of faders provide a graphic representation of the frequency response curve produced by the equalizer.

ground A common signal return point.

half-track A single magnetic audio track that occupies roughly half of the width of a quarter-inch-wide audio tape. See track.

hard disk A permanent recording system based upon rapidly spinning disks having a magnetic coating. Hard disks can be easily erased and rewritten. Also called a hard drive or fixed disk.

hard left; hard right An audio signal placed entirely within the left or right stereo channel.

hardwired Permanently wired in place.

harmonic distortion Degradation of an audio waveform in which distortion products are harmonically related to signal inputs, i.e., at whole number multiples or fractions.

harmonics Multiples or whole number fractions of a fundamental frequency. When combined with a fundamental frequency, harmonics and overtones give a sound its characteristic timbre.

Harmonizer A trade name for Eventide Corporation's pitch-shifting units.

HD Hard disk.

HD recorder A digital recorder that stores recordings on a hard disk.

head drum In a rotary-head recorder, the metallic circular shell containing the rotating head disk or wheel mechanism. The recording tape wraps around the head drum.

head gap In a magnetic recording or playback head, the narrow gap between the poles of the electromagnet. At either side of the gap is one of the electromagnet's ends. During recording, the record head creates a magnetic field around the gap that magnetizes particles on magnetic tape passing through the field. On playback, magnetized tape particles at the gap of the playback head induce an electrical current in the playback head.

headroom A measure of the difference between the highest audio peak level and the level that a piece of equipment can accommodate without signal distortion.

helical scan The path of a spinning head angled across recording tape wrapped in a spiral around a head drum. The result is a series of magnetic tracks angled across the tape. This type of recording method allows tape to head speed to be increased without high-speed tape movement. DAT recorders employ helical scan recording.

hertz A measure of frequency in the number cycles per second. The term honors Heinrich Hertz, the nineteenth-century German physicist who developed the theory of radio waves. Abbreviated as Hz.

high-pass filter A filter that passes audio above a selected frequency while attenuating audio frequencies below that point. Compare with low-pass filter.

horn loudspeaker A highly efficient loudspeaker design having a shape somewhat like that of a trumpet.

hum A steady, low-frequency sound commonly introduced into equipment by poor grounding that allows AC power currents to be picked up by audio circuits.

hypercardioid microphone A microphone with a very narrow, highly directional cardioid pickup pattern. Also known as *super-cardioid.*

hysteresis In magnetic recording, residual magnetism in the iron core of an electromagnetic recording head when current ceases.

I/O Input/Output. A circuit that provides an external interface for a piece of equipment.

IBOC In band on channel. A radio transmission technique in which a station's digital signal shares the same spectrum space as its analog signal.

IEC curve An audiotape equalization standard widely used in Europe.

IEEE 1394 See FireWire.

impedance Impedance is opposition to the movement of AC energy resulting from the net effects of resistance, capacitive reactance, and inductive reactance in an electrical circuit. Measured in ohms.

Inches per second (ips) A measure of the speed of tape movement. On analog tape machines used in broadcasting, tape speeds of 7.5 ips and 15 ips are common. In recording studios that use analog tape decks, speeds of 15 ips and 30 ips are common, with the higher tape speeds yielding improved signal-to-noise ratios.

incidental music In film and video, music used to establish a mood and suggest locale and time era.

induction The creation of EMF as a result of electromagnetic principles, or the creation of magnetic fields through the flow of a current.

infinite baffle A loudspeaker enclosure in which none of the sound emanating from the rear of the cone is allowed to reach the front of the system.

inline consoles Mixers that connect each input channel to a corresponding recorder track. This type of mixer does not have group faders.

intellectual property rights The rights accorded to authors, composers, and inventors to protect benefits due them for their creations.

interaural level difference The difference in acoustic levels experienced at a listener's left and right ears.

interaural time difference The difference in timing of sound waves experienced at a listener's left and right ears.

interference tube Construction used in shotgun mics to achieve highly directional pickup patterns.

intermodulation The generally undesirable phenomenon that causes production of sum and difference frequencies when two or more audio frequencies are present in a circuit.

iPod The trade name for MP3 players developed by Apple Computer.

ISDN Integrated Services Digital Network. A technology for transporting digital audio through telephone lines.

isobaric In audio, the use of two drivers coupled so as to produce greater sound output.

kilohertz One thousand hertz. Abbreviated as kHz.

land In phonographic recordings and CDs, the flat surface between tracks.

laser Light Amplification by Stimulated Emission of Radiation. A device that creates a coherent, in-phase beam of light used in a variety of ways including the recording and playback of compact disks.

lavaliere microphone A small microphone designed to be hung as a necklace around the neck or attached to the lapel of a performer.

LCD Liquid crystal display. A technology that has applications ranging from information screens on calculators and audio recording equipment to displaying images on television sets and computer monitors.

LED Light-emitting diode. A small component used frequently as an indicator.

levels Signal amplitudes. Audio levels are commonly measured with dB or VU meters.

limiter An electronic device that keeps audio peaks from rising above predetermined levels.

line level The level of an audio signal after preamplification, typically 0 dBm.

linear In a straight line.

listener fatigue Physical exhaustion that may result from listening for extended periods of time to high sound levels, especially on loudspeaker designs that do not reproduce low frequencies well. Listening for long periods of time using headphones can also cause listener fatigue.

live Not recorded. In broadcasting, a program that is aired in real-time. Also, refers to a studio having room surfaces that easily reflect sound waves instead of absorbing them.

live assist A broadcast that is partially live and partially automated.

localize to locate the spatial position of a sound source.

log A record of ordered program or recorded elements arranged according to time.

looping In motion pictures, adding or replacing dialogue for the film's sound track in postproduction.

lossless codec A digital compression/decompression scheme in which all compressed data can be retrieved. Most codecs do not fit this description. The advantage of a lossless codec is that the original waveform can be re-created accurately.

lossy codec A digital compression/decompression scheme in which some of the original data measurements cannot be recovered. Heavily compressed audio and video files contain noticeable artifacts of their lossy processing.

loudspeaker A transducer that converts electrical audio signals to sound waves.

low-pass filter A filter that passes audio frequencies below a selected frequency while attenuating audio frequencies above that point. Compare with high-pass filter.

LP Long play. Refers to a long-play vinyl record, normally twelve inches in diameter with the audio encoded as continuous vibrations in the disk's spiral microgrooves. Also called "33" because of its rotational speed of $33\frac{1}{3}$ revolutions per minute.

Mach 1.0 The speed of sound at sea level.

magnetic soundtrack Magnetic recordings of film soundtracks made either on the film containing images or on separate "full-coat" magnetic film. Contrast with optical soundtracks.

magnetic tape A plastic tape to which is bound a thin layer of highly permeable material capable of retaining magnetism after a magnetic field is impressed on it. Used to record various kinds of signals, including audio and video.

magneto-optical A recording technology that combines laser and magnetic recording principles as used in the MiniDisc format. Media may be rewritten many times.

masking Covering or hiding. In audio, masking occurs when a strong signal at a specific frequency overpowers weaker signals at nearby frequencies, rendering weaker signals inaudible.

mass The property of a body that gives it weight.

master A recording from which all subsequent copies are made.

master disk See record-cutting lathe.

mastering Preparation of a final edited and processed version of a recording.

megahertz One million hertz. Abbreviated as MHz.

microgroove Record grooves which are smaller than those used in mass-marketed 78 rpm records, by convention about 20 micrometers. Microgroove recordings were developed in the 1940s and used in 33 and 45 rpm records.

microphone A transducer used to convert acoustic energy to electrical energy.

microwave Radio signals above the UHF frequency band.

mid-range Audio frequencies in the middle portion of the audio frequency spectrum. Loosely speaking, frequencies between 500 Hz and 6 kHz fall into the mid-range.

mid-side (M-S) A stereo microphone technique in which the output of a bidirectional microphone and an omnidirectional microphone located as closely together as possible are combined to yield left and right channels of audio. With good-quality microphones and proper placement, the stereo effect can be excellent.

MIDI Musical instrument digital interface. A standardized system for interconnection of electronic music instruments such as keyboards, drum machines, and synthesizers.

milliwatt A power level of one-thousandth of a watt.

MiniDisc (MD) An optical disk system developed by Sony Corporation for consumer recording and playback of digital audio.

mix To blend together two or more audio sources.

mix down To mix a multitrack recording to a mono, stereo, or surround-sound master.

mixer Devices used to mix audio. See *console*.

modular consoles Mixers that are constructed of sets of independent circuits contained in modules.

modulation The process of impressing a signal such as audio, video, or data on a carrier wave. See *amplitude modulation/frequency modulation*.

molecules The smallest particle of a substance that maintains its unique properties. Molecules are made of an atom or combination of atoms. Two hydrogen atoms combined with one atom of oxygen create one molecule of water structurally represented as H_2O. The properties of water are distinctly different from both hydrogen and oxygen.

monitor In audio production, the loudspeaker systems used by operators to hear audio materials.

mono Short for monaural.

monophonic Sound recorded and reproduced through a single channel. Not stereo. Also monaural.

MOS In motion pictures and video, "Mit Out Sound." Film shot without simultaneous recording of audio. Sounds can be added to the film later in postproduction.

mother disk See record-cutting lathe.

moves Sounds of rustling clothing and scratching.

MP3 MPEG-3. An audio compression codec specified by the Motion Picture Experts Group. The size of audio files can be reduced significantly when compared with "wav" and AIFF recording of the same original audio.

MPEG Motion Picture Experts Group. A professional organization that recommends technical standards for video codecs, among other things.

MRAM An emerging recording technology that is similar to flash storage but uses less power to read and write signals.

mult-box Boxes containing multiple splitters for the connection of microphones to many transformer-isolated outputs. Also called press box.

multi-effects processor A signal processor that integrates a number of temporal, dynamic, and frequency functions in a single piece of equipment.

multiplex Any means of transporting two or more separate signals through a single pathway. In broadcasting, modulating a carrier so that it contains more than one stream or channel of information. The current FM stereo system is technically "FM multiplex stereo," since the FM primary carrier contains both the monaural audio signal and a stereophonic subcarrier.

multitrack In recording, systems that allow the addition or replacement of tracks in synchrony with previously recorded tracks. More than two tracks in audio recording equipment.

mute To silence. Activating mute on an audio console blocks audio from the associated source and releasing the mute function restores its audio. Unlike the solo function, muting removes affected audio from the console's output.

NAB standard equalization curve A standard preemphasis/deemphasis response curve for analog audio tape recording and playback. Developed by the National Association of Broadcasters to improve signal-to-noise ratios of magnetic audio recording.

Nagra A manufacturer of portable reel-to-reel tape machines used especially in film production.

near coincident technique A stereo microphone technique in which two cardioid microphones are placed several inches apart and at an angle of at least 90° to one another.

near field Monitor loudspeakers that are designed to be placed sufficiently close to engineers that reverberations are not a significant part of the sound heard.

negative phase A negative expenditure of energy. In audio, the negative phase is the portion of a sound wave in which air molecules bounce back, or rarefy.

Neutrik Manufacturer of audio and power connectors.

neutron A subatomic particle located within an atom's nucleus that carries no electrical charge.

noise Undesired audio or sounds.

noise floor The level at which weak audio signals are masked by system noise.

noise gate An audio processor that blocks a signal from passing through when its level falls below a preset threshold value.

nondestructive editing A digital production method that preserves original audio files during editing.

normalize A process of adjusting audio levels so that a selection's peak level is raised to an optimum value without altering the relative levels within the piece.

normalled Connected without the use of patch cables. A way of wiring patch panels so that the most common configuration is automatically connected when no jacks are in place. Once a patch cord is inserted, signals can be rerouted to other configurations.

notch filter A filter that can greatly attenuate a narrow band of audio frequencies.

null A point at which a sound level is at one of its lowest values. Directional microphones and directional broadcast transmissions have nulls, or angles at which sound pickup is much reduced.

Nyquist frequency The highest frequency that can be properly recorded when using a specified sampling rate. Generally, one-half the sampling frequency.

octave Eight musical notes above or below a reference note. A step represented by the multiplication or division of a frequency by the number two.

off-axis coloration Irregularities in frequency response of audio that occur with the placement of an audio source within nulls of a directional microphone's pickup pattern.

omnidirectional microphone A microphone that is equally sensitive to sound arriving from all directions.

optical fiber Thin glass fibers used to transport signals coded on light waves. Using laser or LED light sources, optical fiber can be used to transfer audio, video, or data from point to point.

optical soundtracks Sound track recordings that are optically printed onto film alongside images.

ORTF system A variation of the near coincident stereophonic microphone technique developed by the Office de Radiodiffusion Television Française (the French national broadcasting organization).

oscillation   A signal that changes polarity or alternates between maximum and minimum values periodically and regularly.

outboard   A piece of equipment external to the primary system. In studios, outboard signal processors are usually brought within a mix by auxiliary send/return loops.

overdubbing   In multitrack recording, adding new audio material in synchrony with previously recorded tracks.

overmodulation   In AM broadcasting, modulation of the radio carrier at so high a level that clipping of the audio waveform occurs. According to FCC definition, 100 percent modulation in FM broadcasting is defined as ±75 kHz deviation from the assigned carrier frequency.

oversampling   Artificially increasing the number of data points to improve quality of an audio waveform constructed by a digital playback device.

oversaturate   Raising the signal levels of a magnetic recording beyond the point of saturation. See *saturation*.

overtone   Any note harmonically related to a fundamental frequency.

oxide   In magnetic recording, the ferrous material, usually ferrous oxide, bonded to plastic audiotape.

PA   Public address system. A simple sound reinforcement system for use in live venues.

pad   A device used to reduce audio levels by a fixed amount.

pan   Term short for panoramic. Refers to the position of an audio source within a stereophonic field stretching from far left to far right.

paper cone   The sound-producing diaphragm of a loudspeaker driver.

parabolic microphone   A highly directional microphone using a dish-shaped reflector with a pickup element placed at the parabola's focal point. Used to pick up distant sounds such as those at sporting events.

parametric equalizer   An audio equalizer that permits an operator to concurrently adjust parameters of bandwidth, center frequency, and amount of cut or boost.

passive   Not amplified, not powered.

patch bays   Panels containing an array of input and output connectors that permit engineers to route audio signals from selected sources to desired destinations by inserting patch cords into appropriate jacks.

PCM   Pulse code modulation. A type of coding system for digital audio signals.

peak program meter (PPM)   A meter that displays the instantaneous peaks of audio signals.

percent of modulation   The proportion of allowable modulation level expressed as a percentage. See *modulation*.

permeable   The ability of a material to accept and retain a magnetic field.

phantom power   Power supplied by a mixer to a condenser microphone through its cable, typically 48 V.

phase   In repetitive signals, one cycle is represented as a signal that rises above a starting point, falls below it, and finally returns to the starting point. In audio, phase describes the relationship between waves' positions. If positions of two waves coincide with one another, they are in phase. If the peaks of the two waves fall at different points in time they are out of phase.

phono Short for phonograph.

phono cartridge A transducer that converts the mechanical movements of a stylus in grooves on vinyl disks to an electrical output signal.

phonograph From the Greek, "phono" referring to sound and "graph" meaning to write. Commonly called record players.

photoreceptor A device that is sensitive to light and is able to convert variations in light intensity to corresponding variations in electrical current.

pickup pattern A graphic representation of the relative sensitivity of microphones to sounds coming from different angles around the pickup point.

piezoelectric microphones A microphone design incorporating a crystal or ceramic element. When acoustic energy moves the microphone's diaphragm, it compresses or distorts the crystal element, thus generating an electrical current. Piezoelectric microphones have poor frequency response compared with dynamic, ribbon, and condenser microphones and so have little professional use.

pinch roller The rubber or plastic roller that presses audio tape against a spinning capstan, thereby moving the tape from its feed reel past the deck's heads to a take-up reel. The pinch roller, in conjunction with the capstan, provides the primary tape transport force in a tape deck. Torque is applied to the feed and take-up reels only to keep tape stored smoothly on the reels, not to pull the tape from one reel to the other.

ping-ponging See bouncing.

pink noise White noise that has been equalized or filtered so that energy per octave is approximately equal across the audio spectrum. See *white noise*.

pit In CDs a microscopic depression in the surface of the disk used to encode data. See *land*.

pitch A term that refers to the frequency of a musical note. High pitches correspond to treble frequencies, and low pitches refer to bass frequencies.

pitch shifter An audio signal processor that changes the pitch of an audio signal either to compensate for deficiencies in its sound or to achieve a creative effect.

planar magnetic loudspeaker A type of loudspeaker constructed of a long narrow thin metallic ribbon suspended between two permanent magnetic poles. An audio signal applied to the ribbon causes it to be attracted to or repelled by the magnetic fields. Used only in high-frequency drivers.

plasma display A flat panel display technology that uses an electrical discharge to illuminate figures by ionizing the gas cells they contain. Sometimes used in level meters in audio equipment.

plate reverb A way of artificially creating reverberation that became widely used in the 1960s. An audio transducer attached near an edge of a large, tightly suspended metal plate causes sound waves to be set up in the plate, creating simulated reverberation. Pickup devices attached at other points on the plate reconvert sound waves to an electrical signal containing the reverberant sound.

playback head The transducer in a tape deck that converts magnetic field variations contained on recorded magnetic tape to electrical output signals.

podcasting A term applied to the dissemination of programs recorded as MP3 files. Subscribers to podcasting services may have the files automatically downloaded via the Internet to their computers, from which they may then download files to MP3 players. Named for Apple Computer's MP3 player, the iPod.

polar pattern Polar patterns are circular graphic representations of the pickup patterns of microphones showing their sensitivity to sounds arriving from different directions around their pickup points.

polarization A term used to describe the alignment of a magnetic field, that is, north or south.

polyvinyl chloride PVC, a plastic material. Commonly used for manufacture of 33 and 45 rpm vinyl records.

pop filter A foam filter or screen placed between the lips of a vocalist or talent and a microphone. Used to avoid overloading microphones by blasts of air from speech plosives.

POP A term used for a describing a method of processing email. An acronym for "Post Office Protocol."

ported enclosure A loudspeaker enclosure in which sound from the rear of the cone is routed to the front through a hole or port.

positive phase A positive expenditure of energy. In audio, the positive phase is the portion of a sound wave in which air molecules are compressed.

post-equalizer A routing that occurs after processing by the equalizer stage of a channel.

post-fader A routing that occurs after level adjustment by faders of a channel.

postproduction Editing and processing previously recorded finished materials.

potentiometer Often called a "pot." A variable resistor such as a volume or gain control.

POTS Plain old telephone service (or system). Conventional telephone service.

pre-fader A routing that occurs before level adjustment by faders of a channel.

preamplifer An amplifier used to raise low-level signals to a level suitable for input to other devices.

predelay A control parameter of a reverb processor that is used to adjust the amount of early reflections.

preemphasis An equalization process that boosts high frequencies according to standard response curves during recording or broadcast, thus raising those frequencies above noise levels. Upon playback or reception, both the gain of the high frequencies and the unprocessed noise are reduced (deemphasis) in order to achieve proper frequency response. The end result is improved signal-to-noise ratios.

pre-equalizer A routing that occurs before processing by the equalizer stage of a channel.

preproduction The stage of project during which all aspects of a production are planned.

principle of equal loudness The principle that describes human hearing's irregular frequency sensitivity. The human ear is most responsive to frequencies between 1 kHz and 6 kHz and declines sharply below 500 Hz and above 7 kHz. In order to achieve equal loudness, sounds at the extreme ends of the audio spectrum require greater energy. See *Fletcher-Munson curves*.

processors Devices capable of altering one or more parameters of an audio signal.

producer The person responsible for the creation of audio productions or, alternatively, the person responsible for the overall planning and oversight of any audio creation.

production The act of producing a creative work.

proton A positively charged subatomic particle. Protons join neutrons to form the atom's nucleus. The number of protons found in an atom varies from one element to another, as described in the periodic chart of elements.

proximity effect A characteristic of most cardioid microphones that results in a boost of low frequencies when a sound source is close to the mic. Sometimes called "bass-end tip-up."

psychoacoustics The scientific study of the human perception of sound and the interpretation of sounds by human brains. Some of the areas of inquiry in psychoacoustics include localization of sounds, perception of pitch, and interpretation of complex aural signals.

pucks In mastering vinyl recordings, the balls of raw material from which records are pressed.

punch-in/punch-out A technique for rerecording a section of a previously recorded audio track by playing the track, entering record mode at the appropriate moment, and then exiting record mode when the section has been replaced. This leaves the remainder of the track unchanged.

quadraphonic A four-channel audio system consisting of front right, front left, rear right, and rear left.

quantizing Binary measurement of the voltage of digitally sampled waveforms.

quarter-inch connector A single-pin connector, one-quarter inch in diameter. Also known as the phone plug.

quarter-track Tracks one-fourth the width of a quarter-inch tape. Interleaved paired tracks provide stereo recordings, first in one direction on the tape and then in the reverse direction. Quarter-track is primarily a consumer format because of its inadequate signal-to-noise ratios.

R-DAT The term for rotary head digital audiotape. Commonly referred to simply as DAT.

RAM Random access memory. In computers, RAM provides temporary storage for data and for computations.

rarefaction Portions of a sound wave in which air molecules are reduced in density. The negative phase of a sound wave during which molecules "bounce back" after compression.

ratio In a limiter or compressor, the ratio of the change in level at the input (in dB) corresponding to a one dB change in level at the output.

RCA connector A connector made of a single central pin surrounded by a round outer shield.

real-time analyzer (RTA) A device for creating a graphic display of unevenness in frequencies reproduced by a live sound system. Useful for "tuning a room" by means of equalization in order to get the most accurate reproduction of sound in a live performance setting. An RTA can also be used to tune the monitoring environment of control rooms.

record-cutting lathe A machine used to cut grooves in lacquer-coated records that are used for master recordings. The master is used to create a mother, and the mother in turn is used to create stampers that are used to press vinyl records for distribution.

recording head The magnetic transducer, also known as a record head, that converts electrical signals into magnetic fields. As magnetic tape passes over the head, its particles are reoriented into patterns corresponding to the magnetic fields of the recording head. The magnetic fields retained on the tape make a record of the input audio signal.

release time A release time control determines how quickly a compressor or limiter will cease reduction of a signal's amplitude after it falls below the threshold level.

remote trucks Large trucks used in production outside the studio. Within the truck are complete control rooms, recording, and monitoring facilities.

remote In broadcasting, programs originating outside the broadcast studio, for example a live stadium concert or football play-by-play.

resistor An electrical component that resists the flow of electrical current.

resonance In acoustics, the condition in which reinforcement of audio waves at certain frequencies occurs due to standing waves caused by sound reflections. In electronics, resonance is the condition resulting from a balance between inductive and capacitive reactance.

resonator An electronic device employing inductive and capacitive elements that can be made to appear resonant at selected frequencies.

return The routing of a signal back to a mixer after external processing.

reverb time A control parameter of a reverb processor that is used to adjust the length of reverberations.

reverberation Multiple reflections of sound waves not heard as distinct echoes.

rhythm In music, an arrangement of notes that emphasizes their length, timing, and stress.

RIAA curve Audio recording and reproduction response curves for phonograph records developed by the Recording Industry Association of America. Used to enhance signal-to-noise ratios.

ribbon loudspeaker See planar magnetic loudspeaker.

ribbon microphone A microphone in which the pickup element consists of a metal foil strip suspended between powerful magnets. This type of microphone tends to be more delicate than most other designs.

room modes The tendency for enclosed spaces such as rooms to exaggerate certain frequencies through resonance and to suppress others due to phase cancellation. The dimensions of a room determine which frequencies are affected in these ways.

room resonance See *resonance*.

root mean square (RMS) The effective value of a sinusoidal electrical voltage corresponding to 0.707 of the peak value of the AC signal.

rough mix A mix used to approximate a finished mix. Often used to provide temporary mixes early in the basic tracking stage of multitrack productions. See also *scratch vocals*.

rpm Revolutions per minute.

S/PDIF Sony/Philips Digital Interface. A digital audio connector standard developed for consumer audio equipment.

sample A voltage measurement taken of a waveform in the conversion of an analog signal to a digital signal.

samplers Keyboard musical instruments that can record and play back short digital audio files called samples. Usually, samples are made of musical sounds. Each key of the keyboard can be set up to play sampled sounds as musical notes so that the keyboardist may play arpeggios, chords, and scales, just as though playing the sampled instrument.

sampling In digital recording, measurements of an analog audio waveform made at precise and regular intervals. For example, compact disk audio recordings are sampled at 44,100 times per second.

satellite radio Radio broadcasts uplinked to satellites orbiting the globe for retransmission to satellite receivers on earth. Currently satellite radio broadcasts are subscription services marketed to consumers by companies such as Sirius and XM.

saturation In magnetic recording, the point at which all of the magnetic particles on a recording tape have been magnetized.

SCMS Serial copy management system. A technology that makes it impossible to produce a digital copy of a digital copy of a digital audio recording. It is a system advocated by the recording industry for use in consumer digital-recording devices intended to halt potential piracy of commercial recordings.

score In music, the written form of a piece of music.

scoring stage A film studio where music is added to a motion picture in postproduction.

scratch vocal A rough, initial recording of a vocal track that can be used as a reference for musicians during a multitrack recording session.

scrubbing Alternately moving forward and backward while listening to a short segment of recorded audio in order to precisely locate an edit or cue point.

SCSI Small computer system interface. A protocol for transferring data between a computer and SCSI-compatible peripherals. SCSI drives permit a rapid movement of data, making them particularly useful for digital audio and video production and postproduction.

sends Circuits in a mixer placed before the output stage to route signals to external equipment. Sends are often paired with a return circuit to reintroduce processed signals to the mix.

sequencer A device, either a computer workstation or separate item of equipment, that sends, receives, and records MIDI messages.

servo speakers Active loudspeaker systems that employ feedback loops to reduce audio distortion.

SESAC Society of European Stage, Authors, and Composers. A music-licensing organization known primarily for its eclectic catalogue of ethnic, international, and gospel songs. See also ASCAP and BMI.

shelving A filter or equalizer that passes a large band of frequencies while cutting other large bands at the upper or lower portions of the audio spectrum. The resulting response curve somewhat resembles a shelf.

shock mount A device used to attach a microphone to a stand, fish pole, or other fixture. Commonly constructed of a rubber web that intercepts and absorbs vibrations, thus acoustically isolating the mic from mechanical impacts.

shotgun microphone A long microphone with an extremely narrow pickup pattern that allows it to capture distant audio effectively. So named because many have a shape somewhat reminiscent of a shotgun barrel.

sibilance Hissing vocalizations such as "s," "sh," and "zh" sounds.

signal-to-noise ratio (SNR) The difference, expressed in decibels, between the noise floor and the maximum possible level. Used to quantify dynamic range of equipment.

sine In mathematics, a trigonometric function used to study cyclical periodic phenomena. In audio, application of the function produces "sine waves" that graphically represent the waveform of a single pure frequency.

single-point arrays Live sound reinforcement loudspeaker systems made of a single cluster, usually placed above center stage.

sinusoidal Having the qualities of a sine wave.

skip In phonograph playback, jumping grooves causing an abrupt jump forward or backward.

slap-back echo A distinctive effect producing a echo having a short delay time with little additional reverberation. The early Elvis Presley sessions at Sun Records in Memphis often used slap-back echo.

SMPTE Society of Motion Picture and Television Engineers. An organization that has been instrumental in developing technical standards for media such as the SMPTE time code used in audio and video production.

SMTP Simple mail transfer protocol. A protocol commonly used for email.

snake A set of audio cables bundled into a common plastic or rubber cable sheath. Snakes make cable runs neat and tidy—but they do resemble snakes.

solo In audio mixers, a function that allows an operator to hear to individual sound sources without affecting the overall mix. See *mute*.

Sonic Maximizer A proprietary audio processing device manufactured by BBE Sound.

sound bites Brief audio excerpts from recorded remarks of a newsmaker. Mainly used in newscasts.

sound card A computer add-in card containing ADC and DAC circuits to convert analog audio inputs and outputs for internal digital processing and storage.

sound check A pre-performance microphone test that allows console engineers to check levels and proper operation of equipment.

sound effects In film, effects added to soundtracks to lend realism to productions.

sound field The area in which sound waves are present.

sound pressure level A measure of acoustic energy in a sound field. Measured in decibels.

sound reinforcement The technique of raising sound levels in live performances that are not loud enough to be heard well by audiences.

spaced pair A type of stereo microphone system employing two mics spaced as much as twelve feet apart. Also called an A-B stereo microphone technique.

Speakon A type of connector used in connecting large high-quality sound systems. Developed by Neutrik.

specific sound effects Specialized sound effects, such as gunshots, crashes, punches, and squeaks, that are used to accompany unambiguous screen action.

spider In a loudspeaker, the flexible ring-shaped fixture that holds the voice coil in position.

split consoles Mixers that incorporate subgroup faders to provide control over combined sets of channels.

splitter In audio, a device that splits the signal from a microphone, allowing it to be routed to more than one destination, e.g., to an FOH mix and to a monitor mix.

spring reverb A reverberation device constructed around a suspended metal spring. A transducer for conversion of input signals into sound waves is attached at one end of the spring and another transducer attached elsewhere detects the sound waves, converting them into an electrical output signal.

sprocket holes Small rectangular holes along the edge of film, used for gripping film by sprockets that pull film through cameras and projectors.

stampers Reverse-image metal disks that, when heated, are used to press balls of polyvinylchloride (PVC) into finished analog recordings for distribution to consumers.

standing waves Sound waves that reflect off surfaces within an enclosed room in such a way that they combine with other waves. In some areas the combination creates increased acoustic levels, while in other areas it creates reduced levels.

stems Submixes used when multiple versions of a production will be created.

stereo enhancer A signal processor that shifts timing between left and right channels to exaggerate a stereo effect.

stereo imaging The placement of sound sources within the stereo field. An impression of sound location in the space reproduced by stereo signals.

stereo Audio that is recorded and reproduced by two channels that approximate sounds heard by the left and right ears.

stop set A group or set of radio production elements, such as commercials, that are played back-to-back. So called because programming stops for the messages.

stylus A tiny pointed cylinder, usually sapphire or diamond, used to trace grooves in vinyl recordings. Stylus movements are transferred to a phonograph cartridge where an output signal is produced. Colloquially, a phonograph needle.

subgroups Sections of a mixer where selected sets of channel outputs can be combined.

subsonic Acoustic waves having frequencies below 20 Hz, thus below the range of human hearing.

subwoofer A loudspeaker used to reproduce frequencies at the lowest portion of the audio spectrum.

super-cardioid microphone See *hypercardioid microphone.*

surface treatment Alteration of an interior architectural surface so as to change its acoustic properties.

surround sound Originally, the system developed by Dolby Laboratories for film sound using four channels: left, right, center, and surround. Now commonly used to describe any multichannel audio system that creates the illusion of a 360 degree sound field.

sweep An equalizer that boosts a set of frequencies that can be moved or "swept" across the audio spectrum.

sweetening Processing of an audio recording in postproduction to enhance its impression by altering sound parameters.

synchronous In sound for video, locking a soundtrack with its associated images through the use of a system that maintains exact timing, most often SMPTE time code.

synthesizer An electronic device, commonly built as a keyboard instrument, that synthetically creates audio waveforms. Synthesizers often simulate musical instruments, sound effects, or totally new audio creations.

take One recording pass.

take-up reel A plastic or metal reel that neatly spools audio tape after it passes an audio tape machine's head assembly.

tape hiss Noise generated by the random arrangement of magnetic particles on analog recording tape.

tape speed The speed of tape movement measured in inches per second.

TCP/IP Transmission control protocol/Internet protocol. A set of protocols used to link computers via the Internet.

telco A colloquial expression for *Tel*ephone *Co*mpany.

temporal An audio processing effect that is based on time-shifting portions of a signal.

temporal masking The diminished capacity for the human ear to hear a soft sound following a loud sound.

3-to-1 rule When using pairs of mics on a single sound source, place the more distant mic no closer than three times the distance of the closer microphone. This placement helps reduce the effects of phase cancellation.

threshold In limiters and compressors, the adjustable input level at which equipment begins to limit or compress audio signals.

threshold of pain Acousticians define 120 dB as the threshold of pain, or the amplitude at which irreparable or permanent damage to hearing will occur. A sensation of ringing in the ears after exposure to high sound pressure levels indicates that the sound pressure levels were at or near the threshold of pain.

THX A set of standards for movie theater sound systems promulgated to ensure that film audio sounds are much the same in all exhibition locations. Developed by Thomas Holman and Lucasfilm.

timbre The mix of fundamentals, harmonics, and overtones that define the characteristics of a sound. Timbre is what gives a piano, or any other instrument, its distinctive sound qualities.

time code A precise locator code that can be recorded on audio, video, and film tracks to maintain their exact synchronization during editing.

timeline In digital workstations, the panel that displays the production elements spread along a horizontal axis.

track sheet A paper form containing information on each track of a multitrack recording.

track An image of an audio signal impressed on magnetic audio tape in the form of magnetized particles along its length.

transducer A device that converts one form of energy into another. For example, a loudspeaker converts electronic energy into acoustic energy.

transfer rate The speed of data movement between interfaced systems.

transients Sounds and sound components of sharp short duration (nonsustaining) such as ones produced by percussion instruments.

transmission line The cable that routes a transmitter's output to an antenna.

transport The controls on an audio tape machine that provide play, rewind, fast-forward, and stop functions.

treble In audio, frequencies at the upper range of human hearing. In music, generally notes above middle C.

tri-amplification Loudspeaker systems used in live sound reinforcement that use three separate amplifiers for high, mid-range, and low frequencies.

TRS Tip-ring-sleeve. A connector made of a single pin having separate connections at the tip, midway as a ring, and at the base as a sleeve.

TS Tip-sleeve. A connector made of a single pin having separate connections at the tip and at the base as a sleeve.

TT Tiny telephone connector, a single-pin connector similar to, but slightly smaller than, a quarter-inch connector.

tube More correctly vacuum tube. An amplification device that produces gain by action of thermionic processes in which electrons are thrown off by conductors heated to incandescence.

TVRO Television receive only. Receiving apparatus composed of an antenna and receiver intended for the reception of signals relayed via a satellite orbiting the earth.

tweeters Loudspeaker drivers intended for reproduction of frequencies in the upper part of the audio spectrum. Used in multidriver speaker systems.

twisted pair Ordinary two-wire telephone system.

two-track A tape-recording format that places left and right stereo signals on approximately half of a quarter-inch tape width.

two-way Speaker systems that use two drivers—a woofer, and a tweeter—to provide coverage of the complete audio spectrum. Three-way speaker systems employ a tweeter, mid-range, and woofer drivers. See *crossover networks*.

tympanic membrane Anatomical term for ear drum.

ultrasonic Audio frequencies that lie above the range of human hearing.

unidirectional A type of microphone that is most sensitive to sounds arriving from a single direction.

unity gain An active device in which the output is at the same level as the input.

upload To transfer a digital file from a local computer to remote computer or server via a network, such as the Internet.

USB Universal serial bus. A technology that allows USB-compatible peripheral devices to be connected to USB-equipped computers.

variable resistors Resistors whose value may be altered, typically by rotating a control or by sliding a control along a short path.

velocity (of sound) In acoustics, a term used to describe the speed at which sound waves travel; at sea level, approximately 1,130 feet per second. Approximately 770 miles per hour.

vibration A mechanical oscillation.

virtual track An audio track created in the virtual environment of digital workstations, equivalent to real magnetic tracks recorded on audio tape.

voice coil In a speaker, the coil through which current flows producing a magnetic field that moves the cone or diaphragm. In a microphone, the voice coil provides an output current through its movement within a magnetic field.

voltage controlled amplifier (VCA) In audio mixing, a circuit the gain of which is determined by an applied control voltage. Employed in level management stages of automated mixers.

VU meter Volume unit meter. Used to measure audio levels in recording and playback equipment.

watt A measure of power or the rate of doing work.

WAV An audio file type developed for use in storage of digitized audio by Microsoft Windows applications. Uses the file extension .wav.

wave A periodic or cyclical motion or variation in intensity or energy, especially when traveling through a medium such as air.

wavelength The physical distance traveled by an audio wave as it completes one full cycle.

wedges In live sound reinforcement, loudspeaker enclosures, usually wedge-shaped, placed on the floor in front of performers to enable them to hear a mix of the group's performance.

wet In audio production, signals that have been processed.

white noise Audio signal containing all frequencies within the range of human hearing at equal levels. To the ear, appears as a hissing sound.

wild sound Nonsynchronous sound recorded separately from the capturing of images.

windscreen A device used to protect microphones from wind and plosive noise or potential damage. Usually constructed of plastic foam.

woofers Loudspeaker drivers intended for the reproduction of low frequencies in multidriver speaker systems.

word A digital number.

word length The number of data bits that comprise a digital number.

wow Slow, periodic changes in speed during playback of a recording resulting in a repetitive variation in reproduced frequencies. See *flutter*.

X-Y coincident technique A stereo mic system made of a pair of cardioid microphones mounted so that their diaphragms are near to each other, but angled at least 90 degrees from each other.

XLR A three-pin connector widely used in professional audio. Especially used for microphones.

INDEX